ARMY SCIENCE AND TECHNOLOGY FOR HOMELAND SECURITY

REPORT 2
C4ISR

Committee on Army Science and Technology for Homeland Defense—C4ISR
Board on Army Science and Technology
Division on Engineering and Physical Sciences

NATIONAL RESEARCH COUNCIL
OF THE NATIONAL ACADEMIES

THE NATIONAL ACADEMIES PRESS
Washington, D.C.
www.nap.edu

THE NATIONAL ACADEMIES PRESS 500 Fifth Street, N.W. Washington, DC 20001

NOTICE: The project that is the subject of this report was approved by the Governing Board of the National Research Council, whose members are drawn from the councils of the National Academy of Sciences, the National Academy of Engineering, and the Institute of Medicine. The members of the committee responsible for the report were chosen for their special competences and with regard for appropriate balance.

This study was supported by Contract No. DAAD19-02-C-0049 between the National Academy of Sciences and the Department of the Army. Any opinions, findings, conclusions, or recommendations expressed in this publication are those of the author(s) and do not necessarily reflect the views of the organization that provided support for the project.

International Standard Book Number 0-309-09164-0 (Book)
International Standard Book Number 0-309-53071-7 (PDF)

Cover: The Pentagon burning after being struck by a hijacked commercial airliner, September 11, 2001. Courtesy of Reza Marvashti, *The Free Lance-Star,* Fredericksburg, Virginia.

Limited copies are available from:

Board on Army Science and Technology
National Research Council
500 Fifth Street, N.W.
Washington, DC 20001
(202) 334-3118

Additional copies are available from:

The National Academies Press
500 Fifth Street, N.W.
Lockbox 285
Washington, DC 20055
(800) 624-6242 or (202) 334-3313
(in the Washington metropolitan area)
Internet, http://www.nap.edu

Printed in the United States of America

THE NATIONAL ACADEMIES

Advisers to the Nation on Science, Engineering, and Medicine

The **National Academy of Sciences** is a private, nonprofit, self-perpetuating society of distinguished scholars engaged in scientific and engineering research, dedicated to the furtherance of science and technology and to their use for the general welfare. Upon the authority of the charter granted to it by the Congress in 1863, the Academy has a mandate that requires it to advise the federal government on scientific and technical matters. Dr. Bruce M. Alberts is president of the National Academy of Sciences.

The **National Academy of Engineering** was established in 1964, under the charter of the National Academy of Sciences, as a parallel organization of outstanding engineers. It is autonomous in its administration and in the selection of its members, sharing with the National Academy of Sciences the responsibility for advising the federal government. The National Academy of Engineering also sponsors engineering programs aimed at meeting national needs, encourages education and research, and recognizes the superior achievements of engineers. Dr. Wm. A. Wulf is president of the National Academy of Engineering.

The **Institute of Medicine** was established in 1970 by the National Academy of Sciences to secure the services of eminent members of appropriate professions in the examination of policy matters pertaining to the health of the public. The Institute acts under the responsibility given to the National Academy of Sciences by its congressional charter to be an adviser to the federal government and, upon its own initiative, to identify issues of medical care, research, and education. Dr. Harvey V. Fineberg is president of the Institute of Medicine.

The **National Research Council** was organized by the National Academy of Sciences in 1916 to associate the broad community of science and technology with the Academy's purposes of furthering knowledge and advising the federal government. Functioning in accordance with general policies determined by the Academy, the Council has become the principal operating agency of both the National Academy of Sciences and the National Academy of Engineering in providing services to the government, the public, and the scientific and engineering communities. The Council is administered jointly by both Academies and the Institute of Medicine. Dr. Bruce M. Alberts and Dr. Wm. A. Wulf are chair and vice chair, respectively, of the National Research Council.

www.national-academies.org

COMMITTEE ON ARMY SCIENCE AND TECHNOLOGY FOR HOMELAND DEFENSE—C4ISR

JOHN W. LYONS, NAE, *Chair,* U.S. Army Research Laboratory (retired), Mount Airy, Maryland
DENNIS J. REIMER, *Vice Chair,* U.S. Army (retired) and Memorial Institute for the Prevention of Terrorism, Oklahoma City
DUANE A. ADAMS, Carnegie Mellon University, Arlington, Virginia
HENRY L. BERTONI, Polytechnic University, Brooklyn, New York
JAMES J. CARAFANO, The Heritage Foundation, Washington, D.C.
GEORGE M. CLARK, Radiance Technologies, Inc., Huntsville, Alabama
TIMOTHY COFFEY, University of Maryland, College Park, and National Defense University, Washington, D.C.
ANTHONY C. DIRIENZO, COLSA Corporation, Huntsville, Alabama
MITRA DUTTA, University of Illinois, Chicago
FREDERICK L. FROSTIC, Booz Allen Hamilton, McLean, Virginia
C. WILLIAM GEAR, NAE, NEC Research Institute, Princeton, New Jersey
JAMES R. KLUGH, U.S. Army (retired) and Dimensions International, Inc., Alexandria, Virginia
JOSEPH P. MACKIN, E-OIR Measurements, Inc., Spotsylvania, Virginia
LOUIS C. MARQUET, Consultant, Long Branch, New Jersey
LOIS C. McCOY, National Institute for Urban Search and Rescue, Santa Barbara, California
CHANDRA KUMAR N. PATEL, NAE, NAS, University of California at Los Angeles
ALBERT A. SCIARRETTA, CNS Technologies, Inc., Springfield, Virginia
ANNETTE L. SOBEL, Sandia National Laboratories, Albuquerque, New Mexico
MICHAEL F. SPIGELMIRE, U.S. Army (retired) and Consultant, Destin, Florida
LEO YOUNG, NAE, Consultant, Baltimore, Maryland

Liaisons, Board on Army Science and Technology

ROBERT L. CATTOI, Rockwell International (retired), Dallas, Texas
DONALD R. KEITH, U.S. Army (retired) and Cypress International (retired), Alexandria, Virginia

National Research Council Staff

MARGARET N. NOVACK, Study Director
JAMES C. MYSKA, Research Associate
CARTER W. FORD, Senior Project Assistant

Preface

This is the second study in a series of three sponsored by the Deputy Assistant Secretary of the Army for Research and Technology. It was conducted by the Committee on Army Science and Technology for Homeland Defense—C4ISR[1] of the Board on Army Science and Technology in the Division on Engineering and Physical Sciences of the National Research Council. The statement of task for this second report is as follows:

> In this follow-on study, focusing on the C4ISR area and the first responder mission, the National Research Council will:
>
> - examine stated capabilities needed for Homeland Security and the Army's Objective Force,[2] identifying and describing areas in which the two communities have similar technical needs and in which collaboration may be possible.
> - highlight technology and systems solutions under development (in both S&T and Acquisition) for the Objective Force, both in the Department of Defense and commercially, which might meet the needs of the Department of Homeland Security.
> - describe other issues that should be addressed in order to facilitate collaboration and sharing of research.

---

[1]C4ISR is the acronym for command, control, communications, computers, intelligence, surveillance, and reconnaissance.

[2]The Objective Force is now called the Future Force and is referred to as such throughout this report.

- prepare a consensus report documenting the study results and containing findings and recommendations to assist the Army.

FOUNDATION PROVIDED BY THE FIRST STUDY

In September 2001, the U.S. Army asked the Board on Army Science and Technology (BAST) to investigate how science and technology might better enable the Army to accomplish its mission in the homeland. The initial BAST report (completed before the establishment of the new Department of Homeland Security) surveyed a broad range of relevant technologies, recommending that the Army take advantage of potential transferability between technologies for the Future Force and those for homeland security.[3] In the C4ISR area, the committee noted that the Army will need the capability to establish links between its first responder military units and civilian first responders to emergency events. The committee also took the view that the Army should play a major role in providing emergency C4ISR in the event of a major natural or terrorism disaster in which civilian systems are seriously impaired. The committee further concluded that the architecture and technology needed for a C4ISR system for homeland security are compatible with the Army's framework for developing and fielding the Future Force, although the Future Force system would have to be adapted or extended to meet the different mission and challenges of homeland security.

The first report was written in a relatively short period of time. Because of the extensive scope of the review, the lack of a well-defined national operational framework,[4] and the time-sensitive nature of the Army's interest, the committee did not study specific products but rather considered technologies one level above individual products, processes, or services.

COMMITTEE COMPOSITION AND PROCESS FOR THE CURRENT STUDY

The second study began with a review of the membership of the first committee and the nomination to the second committee of members with the necessary expertise in C4ISR. The membership of the Committee on Army Science and Technology for Homeland Defense—C4ISR was chosen to include representation from three communities: the military sector, the emergency responder community, and the C4ISR scientific and technical world. The scientific and technological skill sets of the membership include communications, computer science, sensors and guidance, information science, systems engineering, model-

[3]See National Research Council, Science and Technology for Army Homeland Security: Report 1, The National Academies Press, Washington, D.C., 2003.

[4]National operational framework refers to a plan that the Army would use to conduct whatever operation might be necessary in response to a terrorist attack.

ing and simulation, and systems analysis. Although there is no classified material in this report, a security clearance was considered essential, as many of the topics that would be of interest to the committee are classified.

The committee spent considerable time deliberating on how to address the statement of task. It determined that the report should focus on the response phase of a catastrophic event rather than attempt to consider the prevention of such an event. This approach was justified because the response phase would be the time when most emergency responders would be engaged and when emergency C4ISR capabilities would be most called upon.

The committee also chose not to address commercial items, for a variety of reasons. To begin with, the timing of the study as required by the contract was constrained. Additionally, the Army now uses commercial off-the-shelf (COTS) equipment whenever possible, and the committee believed that whatever COTS items might be of interest would already have been embedded in the Army Future Force technologies. Nevertheless, the committee admits that it may have missed some of the more innovative COTS technologies.[5] Lastly, in order to do justice to a commercial equipment survey, the committee believed that it would have had to review a large variety of products, which could have entailed the requirement to review the claims of multiple vendors for the same products. The committee did not wish to try to distinguish between what was claimed for products and what they could actually deliver, nor did it want to subject itself or the National Academies to criticism for overlooking a particular vendor's product.

The committee held two meetings to familiarize its members with the capabilities required for homeland security and the applicable C4ISR technologies that are available or under development for the Army's Future Force. Two more meetings were devoted to writing and coming to a consensus on the findings, conclusions, and recommendations presented in the report.

As was the case with the first report, even as this report was being prepared doctrine and policy were being developed and amended at all levels of government. The Department of Homeland Security (DHS) and the Department of Defense's (DOD's) Northern Command, which are to have the major responsibilities and authority for homeland security at the national level, had been established and were in the early stages of formation and organization. The actual role that will be played by the Army in homeland security must certainly depend in large measure on the operational assignments that Army units will be given in the framework of, or in support of, these overarching organizations. The details remain in a state of flux. As is indicated in the report, while it is anticipated that much of the doctrine will be drawn from existing protocols, the lack of specific doctrine made the study of specific equipment requirements difficult.

---

[5]For example, the Defense Collaborative Tool Suite, a flexible COTS-based suite of applications software, is endorsed by the Office of the Secretary of Defense and the Joint Staff.

REPORT ORGANIZATION

The introductory chapter provides a context for the rest of the report by describing the government's organization for homeland security, beginning with the DHS, followed by the elements of the DOD that will play a role in homeland security, and lastly, the community of civilian emergency responders. A short section compares the ways in which the DOD and local emergency responders acquire their equipment. The chapter closes with a description of a series of potential scenarios illustrating how complexities will mount as additional events requiring emergency response take place.

Chapter 2 describes how the Army plans to equip the Future Force, drawing attention to certain C4ISR technologies that offer potential for collaborative efforts by the DOD and the DHS. Chapter 3 describes who constitutes the emergency responder community, what they are trying to accomplish, and the kinds of capabilities and training they need; the chapter ends with a description of Project Responder, an independent effort focusing on the status of equipment for emergency responders. Chapter 4 provides a detailed description of a subset of C4ISR technologies for the Future Force that appear to match the requirements of emergency responders. Chapter 5 discusses possible ways of bridging the gap between the Future Force technologies and emergency responder requirements and suggests means to facilitate collaboration between the DOD and the DHS to help specify and meet those requirements. Chapter 6 provides a complete listing of the report's findings, conclusions, and recommendations. Separate appendixes provide additional background information on committee biographies, meeting topics, organization of the U.S. Army, the Army acquisition system, C4ISR capabilities for the Army's Future Force, C4ISR capabilities needed for the civilian emergency responder, and criteria for technology readiness levels.

The committee would like to recognize the assistance given by the emergency responder community and the U.S. Army in providing information and answering questions from the committee. It is likewise grateful for the assistance of NRC staff members Margaret N. Novack, James C. Myska, Carter W. Ford, William E. Campbell, and Dorothy Sawicki in producing this report.

John W. Lyons, *Chair*
Dennis J. Reimer, *Vice Chair*
Committee on Army Science and Technology
for Homeland Defense—C4ISR

Acknowledgment of Reviewers

This report has been reviewed in draft form by individuals chosen for their diverse perspectives and technical expertise, in accordance with procedures approved by the National Research Council's (NRC's) Report Review Committee. The purpose of this independent review is to provide candid and critical comments that will assist the institution in making its published report as sound as possible and to ensure that the report meets institutional standards for objectivity, evidence, and responsiveness to the study charge. The review comments and draft manuscript remain confidential to protect the integrity of the deliberative process. We wish to thank the following individuals for their review of this report:

MG Jack D'Araujo, U.S. Army National Guard (retired), Knoxville, Tennessee
Michael J. Grove, Consultant, Stafford, Virginia
Michael J. Hopmeier, Unconventional Concepts, Inc., Arlington, Virginia
James C. McGroddy, National Academy of Engineering, IBM (retired), Briarcliff Manor, New York
Richard Nowakowski, Raytheon JPS Communications, Chicago, Illinois
Jimmy K. Omura, National Academy of Engineering, Cylink Corporation (retired), San Francisco, California
James Shea, Filtronic Sigteck, Inc., Columbia, Maryland
George F. Sheldon, Institute of Medicine, University of North Carolina, Chapel Hill
Paul N. Stockton, Naval Post Graduate School, Monterey, California
Robert J. Trew, North Carolina State University, Raleigh

Although the reviewers listed above have provided many constructive comments and suggestions, they were not asked to endorse the conclusions or recommendations, nor did they see the final draft of the report before its release. The review of this report was overseen by Alexander H. Flax, Consultant. Appointed by the National Research Council, he was responsible for making certain that an independent examination of this report was carried out in accordance with institutional procedures and that all review comments were carefully considered. Responsibility for the final content of this report rests entirely with the authoring committee and the institution.

Contents

Figures, Tables, and Boxes

FIGURES

TABLES

BOXES

Acronyms

ACTD	Advanced Concept Technology Demonstration
ANVG	Advanced Night Vision Goggle
ARNET	Army Reserve Network
ASD (HLD)	Assistant Secretary of Defense for Homeland Defense
C	communications
C2	command and control
C3	command, control, and communications
C4ISR	command, control, communications, computers, intelligence, surveillance, and reconnaissance
CINC	commander-in-chief
COP	common operational picture
COTS	commercial off-the-shelf
DA	Department of the Army
DARPA	Defense Advanced Research Projects Agency
DASD	Deputy Assistant Secretary of Defense
DDR&E	Director, Defense Research and Engineering
DHS	Department of Homeland Security
DISA	Defense Information Systems Agency
DMSO	Defense Modeling and Simulation Office
DOD	Department of Defense
DTRA	Defense Threat Reduction Agency
DUSD (S&T)	Deputy Undersecretary of Defense for Science and Technology

EPR	Emergency Preparedness and Response (DHS directorate)
FBCB2	Force XXI Battle Command Brigade and Below
FBI	Federal Bureau of Investigation
FCS	Future Combat System(s)
FEMA	Federal Emergency Management Agency
FFW	Future Force Warrior
FOPEN	foliage penetration
GIG	global information grid
GIS	Global Information System
GPS	Global Positioning System
HSARPA	Homeland Security Advanced Research Projects Agency
HSPD	Homeland Security Presidential Directive
IAB	Interagency Board for Equipment Standardization and Interoperability
IR	infrared
ISR	intelligence, surveillance, and reconnaissance
IT	information technology
JBFSA	Joint Blue Force Situational Awareness
JPO	Joint Program Office
JTF-CS	Joint Task Forcce-Civil Support
JTRS	Joint Tactical Radio System
LW	Land Warrior
LW-AC	Land Warrior-advanced capability
LW-IC	Land Warrior-initial capability
LW-SI	Land Warrior-Stryker Interoperable
M&S	modeling and simulation
METL	Mission Essential Task List
MTI	moving target indicator
NBC	nuclear, biological, and chemical
NCO	network-centric operations
NCW	network-centric warfare
NDMS	National Disaster Medical System
NEST	Networked Embedded Systems Technology
NGA	National Geospatial-Intelligence Agency
NGO	nongovernmental organization

NIMS	National Incident Management System
NORTHCOM	U.S. Northern Command
NRC	National Research Council
NRP	National Response Plan
ODP	Office of Domestic Preparedness
ORD	operational requirements document
OSD	Office of the Secretary of Defense
PDA	personal digital assistant
PDASD	Principal Deputy Assistant Secretary of Defense
R&D	research and development
RDEC	Research, Development, and Engineering Center
RDT&E	research, development, testing, and evaluation
RSTA	reconnaissance, surveillance, and target acquisition
SAR	synthetic aperture radar
SCA	software communications architecture
SDR	software-defined radio
SIGINT/EW	signals intelligence/electronic warfare
S&T	science and technology
STO	science and technology objective
TDA	tactical decision aid
TRL	technology readiness level
TSWG	Technical Support Working Group
UAV	unmanned aerial vehicle
UGS	unattended ground sensor
UGV	unmanned ground vehicle
USAF	U.S. Air Force
USAR	U.S. Army Reserve
USN	U.S. Navy
US&T	Undersecretary for Science and Technology
UV	ultraviolet
WIN-T	Warfighter Information Network-Tactical
WMD	weapons of mass destruction

This report is dedicated to

General Donald R. Keith
United States Army (Retired)

for his quiet voice of reason,
his untiring dedication, and his
exemplary efforts toward making
life better and safer for
America and her soldiers.

Executive Summary

The tragic events of September 11, 2001, shattered the relative calm of the early days of the 21st century by introducing large-scale international terrorism to American soil. These attacks brought an increased sense of urgency and new meaning to the Army's top priority of protecting the U.S. homeland.

This report reflects the deliberations of the Committee on Army Science and Technology for Homeland Defense—C4ISR and is the second in a planned series of three reports of studies requested by the Department of the Army to assist it in better preparing for its emerging responsibilities in homeland security and homeland defense.[1] Building on the first National Research Council report in this

[1]The terms "homeland security" and "homeland defense" are frequently used interchangeably, but for the Army the terms have precise meanings. At the time these reports were requested of the National Research Council, the term of choice was "homeland defense"; hence the name of the committee. However, the Army now uses the more inclusive term "homeland security." This new terminology is reflected in the title of this report and throughout these chapters. The following definitions were provided by Gregory J. Bozek, Army War Plans Division, Army Deputy Chief of Staff, G3, in a briefing to the Committee on Army Science and Technology for Homeland Defense, Warrenton, Va., May 15, 2002:

- *Homeland security:* The preparation for, prevention of, deterrence of, preemption of, defense against, and response to threats and aggressions directed towards U.S. territory, sovereignty, domestic population, and infrastructure; as well as crisis management, consequence management, and other domestic civil support.
- *Homeland defense:* The protection of U.S. territory, sovereignty, domestic population, and critical infrastructure against external threats and aggression.
- *Civil support:* Department of Defense support to U.S. civil authorities for domestic emergencies and for designated law enforcement and other activities.

series, *Science and Technology for Army Homeland Security* (NRC, 2003), this second report emphasizes how the Army, through its efforts to field the Future Force (previously called the Objective Force), could assist the Department of Homeland Security (DHS) and emergency responders in their efforts to respond to a catastrophic event. While many aspects of homeland security and homeland defense overlap, an extremely high correlation exists in the area of command, control, communications, computers, intelligence, surveillance, and reconnaissance (C4ISR). The committee believes that C4ISR is a high-payoff capability that offers great return on investment for the nation. The committee acknowledges that this evaluation was accomplished at a fairly high level of abstraction. It is possible that different conclusions might be drawn should a highly detailed examination be conducted.

The committee's individual findings, conclusions, and recommendations are presented in Chapters 1 through 5 and are grouped together in Chapter 6. The overarching recommendation of this study is as follows:

> **Recommendation.** The Department of the Army, in coordination with the Department of Defense, should carry out the following:
>
> - Work with the senior leadership in the Department of Homeland Security (DHS) to put in place and *to institutionalize a process for collaboration and sharing* between the Army and the DHS;
> - Assist the DHS in *establishing the research, development, testing, and evaluation infrastructure* (i.e., an acquisition process, systems engineering discipline, modeling and simulation technologies, and testing and evaluation facilities) *to support the emergency responder community*;
> - Work with the DHS *to find common areas of science and technology collaboration*, starting with the Future Force technologies identified in this report. Central to this effort will be the development of a framework or architecture to enable the integration of these technologies into an effective system of systems; and
> - Work with the DHS *to establish processes for joint*[2] *operations*, including joint training and exercises, shared standards, and interoperable systems.

BACKGROUND

President George W. Bush declared war on global terrorism with a goal of eradicating it from the face of the Earth. In declaring that objective, the president launched the nation on a campaign with two fronts—overseas and at home. The overseas effort is spearheaded by the Department of Defense (DOD), but it

[2] Joint in this application means between civilian and military.

involves all elements of national power—military, economic, diplomatic, and moral. The paradigm for conducting the overseas "homeland defense" phase of this war is well understood. However, at home the situation is much different. As of this writing (March 2004), no coherent planning paradigm or operational model for homeland security yet exists, and although a national operational concept for emergency response is being developed, no fully approved comprehensive framework exists to pull together the efforts of federal, state, and local responders. While much has been done in homeland security, there is much more to accomplish. The foundation of a national operational framework for emergency response involves partnerships—among federal, state, and local levels of government; between the private and public sectors; and between civilian emergency responders and the military, specifically the U.S. Army. This report deals primarily with the latter partnership.

Organizing for Homeland Security

Responsibility for homeland security as a whole has now been assigned to the Department of Homeland Security, and civilian emergency responders find themselves leading the frontline efforts to respond to terrorism on U.S. soil. The Homeland Security Act of 2002 (Public Law [P.L.] 107-296), the public law establishing the DHS, describes the department's mission as follows:

> (1) IN GENERAL.—The primary mission of the Department is to—
> (A) prevent terrorist attacks within the United States;
> (B) reduce the vulnerability of the United States to terrorism;
> (C) minimize the damage, and assist in the recovery, from terrorist attacks that do occur within the United States;
> (D) carry out all functions of entities transferred to the Department, including by acting as a focal point regarding natural and manmade crises and emergency planning;
> (E) ensure that the functions of the agencies and subdivisions within the Department that are not related directly to securing the homeland are not diminished or neglected except by a specific explicit Act of Congress;
> (F) ensure that the overall economic security of the United States is not diminished by efforts, activities, and programs aimed at securing the homeland; and
> (G) monitor connections between illegal drug trafficking and terrorism, coordinate efforts to sever such connections, and otherwise contribute to efforts to interdict illegal drug trafficking. (P.L. 107-296, Sec. 101)

The DHS performs this mission by, among other things,

- Securing U.S. borders, the transportation sector, ports, and critical infrastructure;

- Synthesizing and analyzing homeland security intelligence from multiple sources;
- Coordinating communications with state and local governments, private industry, and the American people about threats and preparedness;
- Coordinating government efforts to protect the American people against bioterrorism and other weapons of mass destruction;
- Helping to train and equip emergency responders; and
- Managing federal emergency response activities.

At the direction of the Homeland Security Presidential Directive 5 (February 28, 2003), the DHS is developing two plans to assist the nation in preparing for a major disaster or terrorist attack: the National Response Plan (NRP) and the National Incident Management System (NIMS). The NIMS will provide an operational framework for implementing the NRP. Both plans are currently in final draft form.[3]

The DOD has a long history of providing Army support to civil authorities. However, since the terrorist attacks against the United States, new emphasis is being placed on this mission. This emphasis has resulted in the establishment of new offices and commands.

The new Assistant Secretary of Defense for Homeland Defense (ASD [HLD]) has responsibility for providing guidance and policy for the DOD to implement the desires of the Congress. The ASD (HLD) is thus a key position in the DOD's overall effort to help eradicate terrorism.

The operational arm of the DOD's efforts to combat terrorism in the homeland is the U.S. Northern Command (NORTHCOM), established October 1, 2002. NORTHCOM's mission is homeland security and civil support, specifically:

- Conduct operations to deter, prevent, and defeat threats and aggression aimed at the United States, its territories, and interests within the assigned area of responsibility; and
- As directed by the President or Secretary of Defense, provide military assistance to civil authorities including consequence management operations. (NORTHCOM, 2003)

NORTHCOM's area of responsibility is the United States, Canada, and Mexico and the land, sea, and aerospace approaches to these countries. NORTHCOM is planning to provide support to any of the more than 79,000 municipalities scattered across the United States that may be subject to any type of disaster, whether man-made or natural. This combatant command continues to mature and to develop strategies.

---

[3]As of March 2004.

The U.S. Army

Within the DOD, the Army is the service with the most experience in providing support to civilian authorities. It has provided the preponderance of support received by civilian authorities for all disasters for many years and has accumulated considerable experience in this area. The Army has a good understanding of what is required to save lives and mitigate damage and possesses important capabilities that could assist emergency responders.

The Army's organizational structure consists of three components: the active Army, the U.S. Army Reserve, and the Army National Guard. The Army National Guard, because of its dual state and federal responsibilities, is ideally suited to lead the homeland security mission for the Army. The National Guard Bureau has already begun this effort with the establishment and deployment of its Civil Support Teams to deal with the aftermath of an event involving weapons of mass destruction. The National Guard's interface at the state level provides a natural bridge from the Army to emergency responders.

Additionally, the institutional part of the Army [4] provides a well-developed and structured research, development, testing, and evaluation (RDT&E) process and infrastructure that could assist emergency responders in developing, testing, and certifying the technologies needed to enhance their capabilities.

The Army's Future Force is designed to utilize network-centric warfare (NCW) capabilities to support the principles of "*See* first, *Understand* first, *Act* first, and *Finish* decisively." This system-of-systems approach could also be applicable to the requirements for emergency responders to *see*, *understand*, and *act* upon the situations that they face. The opportunities for the DOD and the DHS to leverage this approach are considerable.

Emergency Responders

The Homeland Security Act of 2002 defines emergency response providers as including "federal, state, and local public safety, law enforcement, emergency response, emergency medical (including hospital emergency facilities), and related personnel, agencies, and authorities" (P.L. 107-296, Sec. 2(6)). These responders include hazardous materials response teams, urban search and rescue assets, community emergency response teams, antiterrorism units, special weapons and tactics teams, bomb squads, emergency management officials, and municipal agencies and private organizations responsible for transportation, communications, medical services, public health, disaster assistance, public works, and construction. Key responders also include emergency management personnel and political leaders at all levels who make crucial decisions and assessments during a crisis.

---

[4] That is, the part of the Army that is not composed of tactical units; the institutional Army consists primarily of a recruiting command, a training base for individuals, and a logistical system.

Comparison of Acquisition in the Army and the Emergency Responder Community

The ways in which the Army and the emergency responder community acquire technologies in the form of new products, processes, and procedures differ widely. The DOD has a very well developed model for acquisition, with formal procedures and top-to-bottom management. The emergency responders acquire new technology through local city and municipal purchasing agents. The DOD process is controlled by standards of practice and rigorous testing and certification, while the emergency responder community has far fewer formal procedures and sometimes none at all.

The military acquisition process is designed to minimize failure and the attendant loss of life on the battlefield; however, because it is so methodical, it can be too slow for the purposes of many programs. Various ways have been devised to circumvent this problem. Spiral development, for example, is a process developed and refined by the Army to improve current capabilities through technology insertion. It involves fielding these new capabilities with a test unit, testing by that unit, and using the test results for fielding to the entire force. It is particularly suited for enhancing such capabilities as C4ISR. If it is interested in this process, the DHS might consider spiral development as part of a menu from which to choose options that would work for emergency responders.

Unlike the Army, emergency responders have not had a dedicated RDT&E system at their disposal, and many are concerned by the lack of standardization and certification of items that they must purchase. There are, however, several efforts under way on behalf of responders: the Technical Support Working Group (TSWG) coordinates the federal research programs designed to help responders, and the Interagency Board for Equipment Standardization and Interoperability (IAB) is developing agreed-upon standards for emergency responders. As the DHS continues to mature, the development of a more formal RDT&E system for emergency responders will be required.

Scenarios

The committee developed example scenarios described in Chapter 1 as an aid in assisting both the DOD and the DHS in determining requirements and capabilities necessary to structure a system to protect the homeland. The committee believes that scenarios can be a valuable tool to assist in the planning and execution of emergency response to disasters. They are helpful in determining the capabilities needed for emergency responders and can assist in training at all levels. Individuals can be trained in the specific tasks that they need to accomplish when these scenarios are blended into multidisciplinary, all-hazards training. The result can be more coordinated response to emergency situations. Scenarios also help to provide a framework to improve compatibility between emergency responders and the Army.

CAPABILITIES FOR THE ARMY'S FUTURE FORCE

The Army's Future Force[5] is literally the future Army, with transformational changes required in the areas of leader development, acquisition, training, sustainment, and institutional initiatives. As discussed in Chapter 2, central to the Future Force is the concept of network-centric warfare (NCW).

NCW is intended to provide three fundamental capabilities to the warfighter and his or her commander through real-time networking. NCW shifts the emphasis from platform-based to network-based capabilities, thereby generating the following opportunities:

- *All members of the network will have access to all networked resources within established security protocols.* Even single platforms can access all of the resources residing within the network. These resources include the sharing of situational awareness with all members of the network, so that all NCW participants can immediately see the whole battlefield.
- *Networked commanders can make more informed decisions.* The commander of an NCW force is able to see the whole picture from the viewpoint of any member of his or her network. The commander is thus able to understand the entire situation quickly.
- *A networked force can more effectively and efficiently synchronize its assets.* NCW provides commanders with the capability to generate precise warfighting effects at an unprecedented operational tempo, creating conditions for the rapid countering of adversary courses of action. As a result, operations may become more efficient and the conduct of war may change. For example, close air support operations may be significantly reduced by the increased ability to anticipate the need for air support and thus to avoid or minimize situations that involve a time-critical requirement for conducting air operations in close proximity to friendly forces.

The committee anticipates that, just as NCW acts as a force multiplier on the battlefield, network-centric operations (NCO) could benefit homeland security. While some of the capabilities that are being developed for the Future Force might be too complex or expensive for the use of emergency responders, many of the technologies would be very helpful, and the concept of network-centric operations could provide a common framework for these technologies.

---

[5]The Future Force (previously called the Objective Force) is the force that the Army is planning for its future. The Future Combat System (FCS) is envisioned as being one of several yet-to-be-determined systems that will be part of the Future Force.

CAPABILITIES NEEDED FOR EMERGENCY RESPONDERS

Chapter 3 identifies C4ISR requirements for emergency responders. It addresses capabilities currently lacking, as well as future emerging requirements, defining the following: the scope of the responder community, the tasks that this community could be required to perform, the conditions under which these activities might occur, and the characteristics and functionality of appropriate C4ISR technologies, and a description of training and exercise opportunities. Lastly, it ends with a description of Project Responder, an independent effort focusing on the status of equipment for emergency responders.

It concludes that responders and Army forces share many common needs. In addition to individual C4ISR technologies, the committee observes that the Army's network-centric approach to operations could serve emergency responders equally effectively. Such a system could produce significant efficiencies in terms of sharing skills, knowledge, and scarce, high-value assets; building capacity and redundancy in the national emergency response system; and gaining the synergy of providing a common operating picture to all responders. Network-centric systems could be particularly valuable for responding to large-scale or multiple attacks with weapons of mass destruction, in which responders would have to surge capacity quickly, be able to adapt to difficult and chaotic conditions, and respond to unforeseen situations. The value of a network-centric approach suggests that individual emergency responder systems have much to gain from being linked and integrated into a national system of systems. Emergency responders require C4ISR capabilities similar to those enabled by Army technologies, such as the abilities to perform C4ISR in urban environments, to network new capabilities with legacy systems, and to provide protection and redundancy against attacks on responder assets.

DEFENSE TECHNOLOGIES FOR HOMELAND SECURITY

Chapter 4 focuses on the technologies currently being developed in the Army or other DOD components in the area of C4ISR that may have application to the homeland security mission and the needs of emergency responders. It begins with a general discussion of the technical issues associated with C4ISR, primarily from a broad systems perspective. Some broad-based programs and tools at the integrated system level are identified that should be of interest to emergency responders and the DHS.

The choice of "command, control, and computers" as a grouping was made because of the integral nature of decision-making algorithms and software now so prevalent in command-and-control systems, and in fact enabled by the vast data capacity and fast processing made available by today's computers. "Communications" stands by itself as the backbone of any such system. The "intelligence, surveillance, and reconnaissance" aspect of C4ISR is treated as a single entity because of the overlapping technologies underpinning the area. Finally, other programs and activities within the DOD that, although outside the strict C4ISR arena, offer real value to the emergency responder are discussed. For example, the major investment and significant advantages available in the DOD modeling and simulation arena are highlighted.

POTENTIAL FOR COLLABORATION BETWEEN THE ARMY AND THE DEPARTMENT OF HOMELAND SECURITY

Chapter 5 discusses possible ways of bridging the gap between the Future Force technologies and emergency responder requirements and suggests means to facilitate collaboration between the DOD and the DHS to help specify and meet those requirements. Substantial overlap exists in the capabilities required by civilian emergency responders and by the Army. This overlap confirms the potential for collaborative efforts by the DOD and the DHS and the resultant establishment of a conduit for transferring technologies to state and local emergency responders. Table ES-1 highlights examples of potential collaborative efforts for certain technologies and programs that underpin C4ISR for the Army's Future Force.

Additionally, Chapter 5 explores six major opportunities for collaboration:

- *Systems engineering:* An integrated design approach to optimize the synergistic performance of a C4ISR system or systems of systems, so that its functions are executed in the most efficient and effective manner possible;
- *Technology transfer coordination:* The concept of establishing a joint Department of Homeland Security and Army collaboration forum for sharing mutually beneficial technologies and services with emergency responders;
- *Experimentation, testing, and review:* The extensive system of experimentation, testing, and evaluation processes and assets that exist within the DOD and that could be shared with the DHS where practical to avoid cost and duplication of effort;
- *Training programs:* The ability of the Army to assist the DHS in the development and execution of a multidisciplinary, multiechelon, and multihazard training, simulation, and exercise program for emergency responders;
- *Network-centric operations:* "*See* first, *Understand* first, and *Act* first" are network-centric warfare principles that clearly apply to the domain of the emergency responder. The Army might assist the DHS in the implementation of an integrated communications expressway that will facilitate NCO; and

TABLE ES-1 Bridge Between Department of the Army/DOD Science and Technology for the Future Force and Emergency Responder Requirements

Aspect of C4ISR	Future Force Requirements	Leveraged Collaboration
Communications	Networked communications and data systems	Joint Tactical Radio System (JTRS) (Army Acquisition) JTRS Squad Level (Army S&T) Warfighter Information Network-Tactical (WIN-T) (Army Acquisition) Adaptive joint C4ISR node (Army S&T) Mobile network management (Army S&T)
Command, Control, and Computers	Act decisively	Smart Sensor Web (DUSD S&T) C3-on-the-move demonstration (Army S&T) Future command post technologies (Army S&T) Intelligent information technology (DARPA S&T) C2 in complex and urban terrain (Army S&T) Battle space terrain reasoning and awareness (Army S&T) Forecasting, planning, and resource allocation (USN, USAF, Army S&T) Geospatial information integration and generation (Army S&T) Agile Commander (Army S&T) Decision support systems for C2 (USN S&T) Homeland Security/DA ACTD (Army and DHS S&T) Joint Force Blue Force Tracking ACTD (OSD/DISA S&T) Knowledge fusion (Army S&T) FBCB2
Intelligence, Surveillance, and Reconnaissance	Know what the network knows	Smart Sensor Web (DUSD S&T) Land Warrior (Army Acquisition) Objective Force Warrior (Army S&T) Warfighter Physiological Monitoring System, part of Objective Force Warrior (Army S&T) Joint Intelligence, Surveillance, and Reconnaissance ACTD (OSD S&T) Network sensors for the Future Force (Army S&T) Advanced night vision goggles (Army S&T) Long-wave micro-IR sensors (Army S&T) Urban reconnaissance ACTD (OSD and NGA S&T) Network Embedded Systems Technology (DARPA S&T) UAVs/robotics Fusion-based knowledge for the Future Force Family of interoperable operational pictures

Joint Development Collaboration	Emergency Responder Requirements
Joint interoperable communications between DOD and local responders In-building communications and tracking global information grid	Networked communications and data systems
Decision-support tools and algorithms Information aggregation, fusion, and sorting Intelligence data dissemination to uncleared entities (soldiers or local responders) C4ISR interfaces for simulations	Informed event management
Joint development of chemical/biological/nuclear sensors Smart sensor networks for urban environments Low-cost, disposable, networked, multiphenomenology sensors Urban UAVs and robotics Space, airborne, and terrestrial sensors	Common operational picture

continued

TABLE ES-1 Continued

Aspect of C4ISR	Future Force Requirements	Leveraged Collaboration
Other	Other DOD assets	Joint Virtual Battlespace (Army S&T) Effects of Weapons Simulations (DTRA S&T) Flexible Asymmetric Simulation Toolkit (DMSO and USAF S&T) Joint Conflict and Tactical Simulation-Laser Project (DMSO S&T) Dynamic mission readiness training (Army and USAF S&T) Chemical and biological hazard environment prediction (USN S&T) Portable and mobile power (Army S&T)

NOTES: S&T, science and technology; DUSD S&T, Deputy Undersecretary of Defense for Science and Technology; C3, command, control, and computers; DA, Department of the Army; DARPA, Defense Advanced Research Projects Agency; C2, command and control; USN, U.S. Navy; USAF, U.S. Air Force; ACTD, Advanced Concept Technology Demonstration; OSD/DISA, Office of the Secretary of Defense/Defense Information Systems Agency; FBCB2, Force XXI Battle Command Brigade and Below; IR, infrared; NGA, National Geospatial-Intelligence Agency; UAVs, unmanned aerial vehicles; DTRA, Defense Threat Reduction Agency; DMSO, Defense Modeling and Simulation Office.

- *Standardization efforts:* The Army's Mission Essential Task List training model could be beneficial if adopted by emergency responders. It would allow for definition of the capabilities required to respond to a terrorist attack and would highlight equipment, technology, and training deficiencies. Additionally, common product standards and conformity testing would ensure better interoperability between DOD and emergency responder equipment.

The committee believes that significant opportunities exist for collaboration between the Army and the DHS to secure the homeland more effectively. Several elements of C4ISR could be used to establish a more formal paradigm to address additional areas. The committee also recognizes that the DHS, because of its relative newness, faces numerous organizational challenges. These challenges, coupled with the current high operational tempo of the Army, make policy commitment on behalf of both organizations possibly the biggest hurdle to overcome.

Joint Development Collaboration	Emergency Responder Requirements
Virtual emergency exercises Plume and fire simulators	Other

However, the events of September 11 underscore why all obstacles to this collaborative process must be overcome.

The requirement for C4ISR is ubiquitous, whether for the Army's Future Force or for the future emergency responder. The committee is convinced that quick action on the part of the Army can provide beneficial C4ISR solutions to the Department of Homeland Security that will ensure a high level of interoperability between emergency responders and the Army should our nation be forced again to respond to a catastrophic event on U.S. soil.

REFERENCES

NORTHCOM (Northern Command). 2003. Who We Are—Mission. Available online at <http://www.northcom.mil/index.cfm?fuseaction=s.who_mission>. Accessed March 26, 2004.

NRC (National Research Council). 2003. Science and Technology for Army Homeland Security: Report 1. Washington, D.C.: The National Academies Press.

1

Introduction

This report reflects the deliberations of the second of the three study committees to be convened in response to a request from the Department of the Army to assist it in better preparing for its emerging responsibilities in the realm of homeland security and homeland defense.[1] The first report, *Science and Technology for Army Homeland Security*, was a broad survey of many different types of relevant technologies with possible application for both the Army's Future Force[2] and emergency responders (NRC, 2003).

---

[1]The terms "homeland security" and "homeland defense" are frequently used interchangeably, but for the Army the terms have precise meanings. At the time these reports were requested of the National Research Council, the term of choice was "homeland defense"; hence the name of the committee. However, the Army now uses the more inclusive term "homeland security." This new terminology is reflected in the title of this report and throughout these chapters. The following definitions were provided by Gregory J. Bozek, Army War Plans Division, Army Deputy Chief of Staff, G3, in a briefing to the Committee on Army Science and Technology for Homeland Defense, Warrenton, Va., May 15, 2002:

- *Homeland security:* The preparation for, prevention of, deterrence of, preemption of, defense against, and response to threats and aggressions directed towards U.S. territory, sovereignty, domestic population, and infrastructure; as well as crisis management, consequence management, and other domestic civil support.
- *Homeland defense:* The protection of U.S. territory, sovereignty, domestic population, and critical infrastructure against external threats and aggression.
- *Civil support:* Department of Defense support to U.S. civil authorities for domestic emergencies and for designated law enforcement and other activities.

[2]"Future Force" is the current term for the Army of the future; it was previously called the Objective Force. The Future Combat System (FCS) is envisioned as being one of several yet-to-be-determined systems that will be part of the Future Force.

This second report builds on the previous effort and focuses specifically on capabilities in command, control, communications, computers, intelligence, surveillance, and reconnaissance (C4ISR) being developed for the Future Force. The report compares and contrasts these capabilities with the C4ISR capabilities needed by civilian emergency responders for a catastrophic event. For this study the committee evaluated those capabilities common to both the Future Force and the emergency responder community and identified some of the most likely enabling technologies for the latter. The committee acknowledges that this evaluation was accomplished at a fairly high level of abstraction. It is possible that different conclusions might be drawn should a highly detailed examination be conducted. The report suggests areas in which the Army could develop collaborative efforts with the emergency responder community through the federal Department of Homeland Security (DHS) in order to accomplish the transfer of the most promising Army technologies to that community.

This chapter provides a context for the rest of the report by describing the government's organization for homeland security, beginning with the DHS, followed by the elements of the Department of Defense (DOD) that will play a role in homeland security, and lastly, the community of civilian emergency responders. A short section compares the ways in which the DOD and local emergency responders acquire their equipment. The chapter closes with a description of a series of potential scenarios illustrating how complexities will mount as additional events requiring emergency response take place.

BACKGROUND

The tragic events of September 11, 2001, have been called a defining moment in our history. Clearly, what happened that day dramatically changed this country and how we as citizens live our daily lives. The words of Abraham Lincoln describing another critical period in U.S. history come to mind: "The occasion is piled high with great difficulty and we must rise with the occasion. As our case is new, so must we think anew and act anew" (President Lincoln's Second Annual Message to Congress, December 1, 1862).

President George Bush declared war on global terrorism with a goal of eradicating it from the face of the Earth. In declaring that objective, the president launched the nation on a two-front campaign—overseas and at home. The overseas effort is spearheaded by the DOD, but it involves all elements of national power—military, economic, diplomatic, and moral. The structure to prosecute this campaign is in place. Much planning and training has gone into developing a world-class military force, and the paradigm for conducting the overseas "homeland defense" phase of this war is well understood.

However, at home the situation is much different. The last serious external military threat to the continental United States was in 1812. As this report is being written, despite the establishment of the DHS, no coherent planning para-

BOX 1-1
Some U.S. Agencies and Organizations Involved in Emergency Response

Government Agencies

Federal

- Department of Homeland Security (new)
  —Directorate of Emergency Preparedness and Response
  —U.S. Coast Guard
- Department of Defense
  —Assistant Secretary of Defense for Homeland Defense (new)
  —U.S. Northern Command (new)
  —Departments of the Army, Navy, and Air Force

State

- State Offices of Emergency Preparedness
- State Police
- National Guard

Local (counties, cities, municipalities)

- Fire Departments
- Local Police
- Emergency Medical Services
- Public Health Departments

Nongovernmental Organizations, Private Sector

- American Red Cross
- Utilities
- Churches
- Private Ambulance Services

digm for homeland security yet exists, and although a national operational concept for emergency response is being developed,[3] no approved comprehensive framework exists to pull together the efforts of federal, state, and local responders. While much has been done in homeland security, there is still more to accomplish.

The foundation of a national operational framework for emergency response involves partnership—among federal, state, and local levels of government; between the private and public sectors and between civilian emergency responders and the military. This partnership involves some agencies established as a direct result of the events of September 11 and others with long experience in responding to natural and man-made disasters. Some of these emergency response agencies and organizations are listed in Box 1-1.

[3] See the discussion on the National Response Plan in the following section.

ORGANIZING FOR HOMELAND SECURITY

To enhance understanding of the technologies described in subsequent chapters, the committee describes in the next three subsections how the nation is organized for homeland security and looks briefly at the structure of the U.S. Army and the emergency responder community with which it will be working in the event of a disaster.

Department of Homeland Security

Responsibility for homeland security has now been assigned to the Department of Homeland Security, which was established by the Homeland Security Act of 2002 (Public Law [P.L.] 107-296) and Executive Order 13284 of January 24, 2003. Figure 1-1 provides the department's organizational chart (as of March 2003).

The public law establishing the DHS describes the department's mission as follows:

> (1) IN GENERAL.—The primary mission of the Department is to—
> (A) prevent terrorist attacks within the United States;
> (B) reduce the vulnerability of the United States to terrorism;
> (C) minimize the damage, and assist in the recovery, from terrorist attacks that do occur within the United States;
> (D) carry out all functions of entities transferred to the Department, including by acting as a focal point regarding natural and manmade crises and emergency planning;
> (E) ensure that the functions of the agencies and subdivisions within the Department that are not related directly to securing the homeland are not diminished or neglected except by a specific explicit Act of Congress;
> (F) ensure that the overall economic security of the United States is not diminished by efforts, activities, and programs aimed at securing the homeland; and
> (G) monitor connections between illegal drug trafficking and terrorism, coordinate efforts to sever such connections, and otherwise contribute to efforts to interdict illegal drug trafficking. (P.L. 107-296, Sec. 101)

The DHS performs this mission by, among other things,

- Securing our borders, transportation sector, ports, and critical infrastructure;
- Synthesizing and analyzing homeland security intelligence from multiple sources;
- Coordinating communications with state and local governments, private industry, and the American people about threats and preparedness;
- Coordinating government efforts to protect the American people against bioterrorism and other attacks with weapons of mass destruction;
- Helping to train and equip emergency responders; and

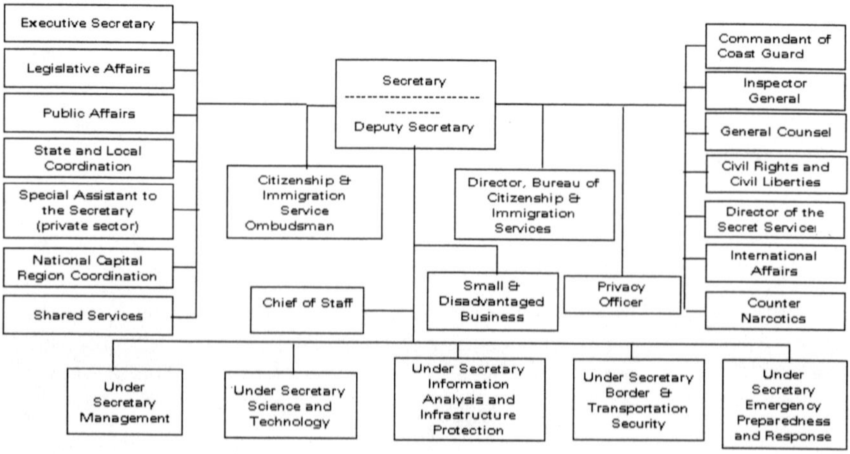

FIGURE 1-1 Organizational chart of the Department of Homeland Security as of March 1, 2003.

- Managing federal emergency response activities.

Within the DHS, the Directorate of Emergency Preparedness and Response (EPR) is of critical importance to the emergency responder community. The newly organized EPR incorporates significant federal emergency responder elements, such as the Federal Emergency Management Agency (FEMA) and response teams from the Department of Health and Human Services (DHHS)[4] and emergency response teams from the Department of Energy. These latter elements were placed under EPR/FEMA specifically to respond more efficiently to the threat scenarios postulated by the DHS in the post-September 11 environment.

Initial National Response Plan

Under the Homeland Security Presidential Directive 5 (HSPD-5, February 28, 2003), the DHS is to prepare the National Response Plan (NRP) to replace the existing Federal Response Plan (see NRC, 2003, p. 157). The NRP is intended to provide an all-discipline, all-hazards approach to domestic incident management and to help federal, state, and local governments work together. The initial NRP was circulated in September 2003. The Secretary of DHS indicated that this initial document implements, on an interim basis, the domestic incident manage-

[4]Most notably, those of the National Disaster Medical System (NDMS) and the Pharmaceutical Stockpile.

ment authorities, roles, and responsibilities of his office, as defined by HSPD-5, until the full NRP becomes effective.[5]

National Incident Management System

HSPD-5 also required the development of a National Incident Management System (NIMS) as part of the NRP. The NIMS will provide an operational framework for implementing the NRP. A draft NIMS was produced in September 2003. A second draft was published March 1, 2004. Compliance with certain aspects of the NIMS is now possible, while other aspects of the NIMS will require further development and refinement.[6]

Department of Defense

The Department of Defense has a long history of providing military support to civil authorities. However, since the terrorist attacks against the United States, new emphasis is being placed on this mission. This focus has resulted in the establishment of new offices and commands. However, many homeland security issues remain unresolved for the DOD; currently the department is focusing on the following matters:

- The evolution of a national vision of military support to civilian authorities,
- The role of the National Guard in homeland security,
- The role of the Coast Guard in homeland security,
- The role of the national laboratories in homeland security, and
- DOD direct support to the DHS.

Assistant Secretary of Defense for Homeland Defense

The National Defense Authorization Act for Fiscal Year 2003 (P.L. 107-314, Sec. 902(a)) established the position of Assistant Secretary of Defense for Homeland Defense with the following mission: "Overall supervision of the homeland security activities of the Department of Defense."

This mission includes the following tasks:

[5]Memorandum from Tom Ridge, Secretary of Homeland Security, to Cabinet Secretaries; Agency Directors; Members of Congress; Governors; Mayors; County, Township and Parish Officials; State Homeland Security Advisors; Homeland Security Advisory Council; and State, Territorial, Local and Tribal First Responders; Subject: Initial Response Plan, September 20, 2003.

[6]Memorandum from Tom Ridge, Secretary of Homeland Security, to Cabinet Secretaries; Agency Directors; Members of Congress; Governors; Mayors; County, Township and Parish Officials; State Homeland Security Advisors; Homeland Security Advisory Council; and State, Territorial, Local and Tribal First Responders; Subject: National Incident Management System, March 1, 2004.

- Overall supervision of the homeland defense activities of the Department of Defense
- Develop strategic planning guidance for DOD's role in homeland security
- Develop homeland defense force employment policy and guidance
- Supervise DOD preparedness activities to support civil authorities in domestic emergencies
- Assist civil authorities in building and improving federal, state, and local homeland security response capabilities
- Plan, train, and perform DOD domestic incident management
- Advocate homeland defense requirements within the Department's resource allocation process. (Cohen, 2003, p. 9)

Figure 1-2 provides the organizational chart for the Office of the Assistant Secretary of Defense for Homeland Defense. An important new responsibility for the assistant secretary is found in the current draft of the National Defense Authorization Act for Fiscal Year 2003 (P.L. 107-314), in which his office is made responsible for overseeing the future technology transfer from the DOD to the DHS. This new responsibility will create a critical channel for the assistance recommended in this report.

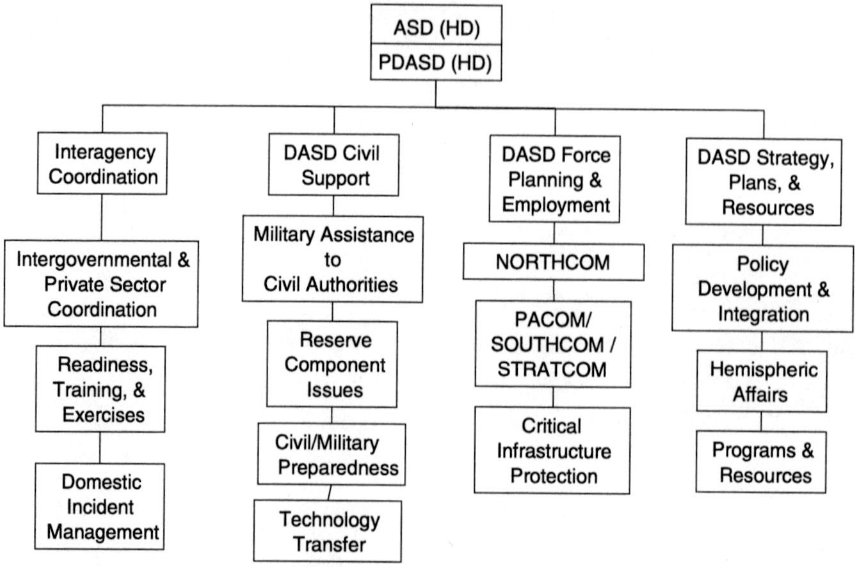

FIGURE 1-2 Organizational chart for the Office of the Assistant Secretary of Defense for Homeland Defense (ASD [HD]). SOURCE: From Homeland Defense, briefing to the committee by Peter F. Verga, Principal Deputy Assistant Secretary of Defense (Homeland Defense) (PDASD [HD]), July 17, 2003.

U.S. Northern Command

The DOD established the U.S. Northern Command (NORTHCOM) effective October 1, 2002. NORTHCOM's mission is homeland security and civil support, specifically:

- Conduct operations to deter, prevent, and defeat threats and aggression aimed at the United States, its territories, and interests within the assigned area of responsibility; and
- As directed by the President or Secretary of Defense, provide military assistance to civil authorities including consequence management operations. (NORTHCOM, 2003a)

NORTHCOM's area of responsibility is the United States, Canada, and Mexico and the land, sea, and aerospace approaches to these countries. Figure 1-3 portrays NORTHCOM's command-and-control relationships. The only forces assigned directly to NORTHCOM are the Joint Force Headquarters for Homeland Security in Norfolk, Virginia; the Joint Task Force–Civil Support (JTF–CS) at Fort Monroe in Hampton, Virginia; and Joint Task Force-6 (JTF-6) at Biggs Army Airfield, Fort Bliss, Texas. Other forces are assigned as needed (NORTHCOM, 2003b). The committee notes that NORTHCOM's organization and missions are in a formative and transitional stage and could be altered in the future.

The U.S. Army

The U.S. Army's mission is to fight and win our nation's wars by providing prompt, sustained land dominance across the full range of military operations and spectrum of conflict in support of combatant commanders. (See Appendix C for information about the organizational structure of the Army.) It accomplishes this mission by carrying out the following:

- Executing Title 10 and Title 32[7] United States Code directives, to include organizing, equipping, and training forces for the conduct of prompt and sustained combat operations on land.
- Accomplishing missions assigned by the President, Secretary of Defense and combatant commanders, and transforming for the future (U.S. Army, 2003).

The U.S. Army consists of three components—the active Army, the U.S. Army Reserve, and the Army National Guard—each of which brings different

---

[7]Title 10 of the United States Code provides for the organization, training, and equipping of all the U.S. Armed Forces, to include the Reserve Components. Title 32 of the United States Code provides for the function of the National Guard while under the control of the state governor. At this point it is not a federal force and is not governed by the Posse Comitatus Act (18 USC 1385). Missions in this status can include crisis management, consequence management, and combatant commander support.

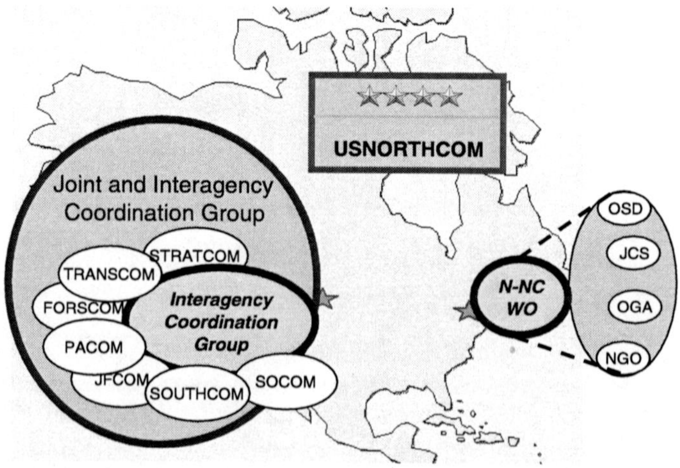

FIGURE 1-3 NORTHCOM command-and-control relationships. NOTE: STRATCOM, U.S. Strategic Command; TRANSCOM, U.S. Transportation Command; FORSCOM, U.S. Forces Command; PACOM, U.S. Pacific Command; JFCOM, U.S. Joint Forces Command; SOUTHCOM, U.S. Southern Command; SOCOM, U.S. Special Operations Command; N-NC, NORAD-NORTHCOM; WO, Washington Office; OSD, Office of the Secretary of Defense; JCS, Joint Chiefs of Staff; OGA, Other Government Agency; and NGO, nongovernmental organization. SOURCE: Stover James, briefing to the NORAD (North American Air Defense Command)/USNORTHCOM Joint and Interagency Coordination Group, August 6, 2003.

strengths to the overall force. The active Army and the Army Reserve provide support to civil authorities in a variety of ways but are generally constrained by law from performing policing duties within the homeland (DHS, 2003). The National Guard—the states' militia—is not constrained by the Posse Comitatus Act (10 USC 1385)[8] as long as it is operating under control of the state governors (Title 32, USC). The National Guard is local in nature and is widely dispersed throughout the homeland. It can be quickly activated by the state governors and can help local police and other emergency responders as required. One would expect that in the aftermath of a large terrorist attack the National Guard would be functioning alongside the local fire, police, and medical personnel. In making

[8]The Posse Comitatus Act of 1878 (18 USC 1385), as a general matter, prevents the Army and the Air Force from directly engaging in law enforcement activities such as search, seizure, arrest, and similar actions.

BOX 1-2
Findings from Report 1 Relevant to the Current Report

Finding 1-1. Homeland security is an important extension of the Army's historical role of providing military support to civilian authorities. The Army will be called on to assist the lead federal agency, the Department of Homeland Security, in meeting a wide range of demands for consequence management and recovery of public order and critical services.

Finding 1-2. The Army National Guard, given its historical mission and flexibility, geographic dispersion, dual-mission capabilities, and frequent association with local agencies, is the key Army asset to meet homeland security demands and can be augmented as necessary with special capabilities from the Army Reserve and the active Army.

Finding 1-3. There are many similarities between military operations involving allied or coalition forces and operations involving civilian emergency responders.

SOURCE: NRC (2003), pp. 24, 29, and 31.

technology packages available to emergency responders, the Army must consider the National Guard as well. The National Guard has anticipated this role. Among several forward-looking steps, it has reorganized its state headquarters into multiservice entities and established Civil Support Teams to respond to the possible use of weapons of mass destruction.

Military support must be capable of smoothly, quickly, and efficiently augmenting the emergency responders in a crisis situation. This kind of assistance implies not only compatible equipment but also commonly understood doctrine and standards as well as joint[9] training to refine operating procedures. Three findings presented in the first report in this series are also relevant to the present discussion (see Box 1-2).

A corollary of the first report's Finding 1-3 concerning the many similarities between military operations involving allied or coalition forces and operations involving civilian emergency responders would indicate that these similarities make it advantageous to consider technologies appropriate to both groups. Indeed, the underpinnings of the network-centric warfare capability envisioned for the Army's Future Force (discussed in Chapter 2 of this report)—"*See* first, *Understand* first, *Act* first, and *Finish* decisively"—correspond directly to the emer-

[9]Joint in this application means between civilian and military.

gency responder's need to *see, understand*, and *act* in order to save lives and mitigate damage caused by man-made or natural disasters.

Emergency Responders

The Homeland Security Act of 2002 defines emergency response providers as including "federal, state, and local public safety, law enforcement, emergency response, emergency medical (including hospital emergency facilities), and related personnel, agencies, and authorities" (P.L. 107-296, Sec. 2(6)). These responders include hazardous materials response teams, urban search and rescue assets, community emergency response teams, antiterrorism units, special weapons and tactics teams, bomb squads, emergency management officials, and municipal agencies and private organizations responsible for transportation, communications, medical services, public health, disaster assistance, public works, and construction. Key responders also include emergency management personnel and political leaders at all levels who make crucial decisions and assessments during a crisis.

While the emergency response needs of fire, police, and emergency medical personnel are receiving considerable attention and increased funding, the critical requirements of other support groups are not as well understood (Jackson et al., 2002). For example, both public health systems and national urban search and rescue assets are widely regarded as essential to emergency response, yet both lack sufficient capabilities to respond to large national emergencies, and little attention has been given to how additional C4ISR capabilities might be used to expand their capacity or improve their efficiency (CFR, 2003).

The needs of responders in the private sector have received even less attention. For example, in the wake of the September 11 attacks, the World Trade Center site required about 10,000 skilled support personnel (heavy equipment operators, truck drivers, iron workers, carpenters, and laborers) per day during the initial search and cleanup period (CFR, 2003). Their operations were essential to the response and entailed significant health and safety risks that could have been mitigated by better C4ISR capabilities at the incident site (Lippy and Murray, 2002).

Another category of resources frequently overlooked in needs assessments is that of the response assets required to deal with agricultural emergencies that either threaten the U.S. food supply or are potential sources of human infectious disease. Animal diseases, for example, can present a serious risk to humans. Many diseases can infect multiple hosts. Three-quarters of emerging human pathogens are zoonotic (i.e., they can be readily transmitted back and forth between humans, domesticated animals, and wildlife). Whether or not they infect humans, animal diseases can have fearful economic impact. For example, Great Britain's response to the 2001 outbreak of foot-and-mouth disease, including lost productivity, amounted to $11.6 billion (Matthews and Buzby, 2001).

While infectious disease is an ever-present danger in a globalized world, the possibility of terrorists intentionally introducing vectors or bacteria or viruses into a population to foster the spread of disease introduces an added dimension to the danger. Thus, agricultural response assets could well be an important component of the consequence management system required to meet the threat of terrorist attacks, and these response assets could have significant C4ISR needs. The Federal Emergency Management Agency has identified this area as being in need of improvement (FEMA, 1997).

Finally, in addition to the state and local assets already mentioned, such as the Army National Guard Civil Support Teams,[10] and private sector assistance, emergency responses could involve a range of federal capabilities. These would include the active forces from the Army, Air Force, Marine Corps, and Navy, the Reserve forces from all services, as well as a range of federal response teams such as Domestic Emergency Support Teams, Disaster Medical Assistance Teams, Coast Guard National Strike Teams, and Nuclear Incident Response Teams.[11] The needs of these various groups and their capacities to integrate into the overall national response system also require consideration.

> **Finding 1-1.** Although a number of informal mechanisms exist, no coherent planning paradigm for the interface between the military and the emergency responders currently exists, and although a national operational concept for emergency response is being developed, it is not yet a comprehensive framework that pulls together the efforts of federal, state, and local responders.

Indeed, the committee conducting the first study in this series reached the conclusion shown in Box 1-3.

COMPARISON OF ACQUISITION IN THE ARMY AND IN THE EMERGENCY RESPONDER COMMUNITY

The ways in which the Army and emergency responder community acquire technologies in the form of new products, processes, and procedures differ widely. The DOD has a very well developed process for acquisition, with formal procedures and top-to-bottom management. Emergency responders acquire new technology through local city and town purchasing agents. The DOD process is controlled by standards of practice and rigorous testing and certification, while the emergency responder community has far fewer formal procedures and sometimes none at all.

---

[10]National Guard Civil Support Teams are not exclusively state and local assets. These organizations could be brought into active federal service under Title 10 of the United States Code. In that status, they could be employed as part of NORTHCOM or any other military command structure as deemed appropriate to meeting the DOD requirement for homeland security.

[11]Michael Lowder, Department of Homeland Security, Federal Emergency Management Agency, Response Division, briefing to the committee, Washington, D.C., August 25, 2003.

BOX 1-3
Conclusion and Recommendation from Report 1 Relevant to the Current Report

Conclusion 4-1. A new national emergency response command, control, and communications system for homeland security must be developed and fielded to meet the demands of the emerging threats, particularly to integrate the response to chemical, biological, high explosive, radiological, and nuclear weapons. This system must be compatible with developments in the new Department of Homeland Security, the U.S. Northern Command, and state and local entities. Current Army science and technology thrusts and programs that are integral to the Objective Force can be adapted for the new national system.

Recommendation 4-1. To facilitate the development and fielding of an integrated command-and-control system for homeland security, the Army should initiate or continue research that permits the earliest possible fielding of deployable communications packages equipped with universal multiplexer capability to facilitate command and control across the vast, and disparate, array of agencies that will respond to incidents and events.

NOTE: Universal multiplexer capability refers to the broad capability for handling several different types of datastreams at an interface where they can be periodically sampled.

SOURCE: NRC (2003), p. 97.

The military acquisition process is designed to minimize failure and the attendant loss of life on the battlefield; however, because it is so methodical, it can be too slow for the purposes of many programs. Various ways have been devised to circumvent this problem. Spiral development, for example, is a process developed and refined by the Army to improve current capabilities through technology insertion. It involves fielding these new capabilities with a test unit, testing by that unit, and using the test results for fielding to the entire force. It is particularly suited for enhancing such capabilities as C4ISR. If it is interested in this process, the DHS might consider spiral development as part of a menu from which to choose options that would work for emergency responders.

The Army acquisition process (see Appendix D for details) begins with the definition of needed capabilities, which are spelled out in a requirements document. The Army science and technology (S&T) arm, consisting of laboratories,[12] engineering centers, and grants-in-aid and contracting offices, responds to the requirements document with technical programs that move through a series of prescribed stages until technologies transition to demonstration and validation efforts and eventually to prototype systems. In this process, the Army S&T

community is increasingly using formal cooperative alliances with private sector entities—universities and industry laboratories—to make use of the best expertise available, wherever it may be found. Once a technology is successfully demonstrated through developmental testing, mature products are transitioned to acquisition program managers for integration into systems or system-of-systems procurement efforts. Testing occurs during the development phase and operational testing is done in the early fielding stages, the latter with troops using the technology in simulated missions. The spiral development process greatly reduces the time associated with technology insertion and has great promise for resolving the challenges faced by emergency responders, particularly in the area of C4ISR.

In contrast, there is little of this type of formal process available to emergency responders. They have not had a dedicated research, development, testing, and evaluation (RDT&E) system at their disposal, and many are concerned by the lack of standardization and certification of items that they must purchase. There are, however, several efforts under way on behalf of emergency responders: the Technical Support Working Group (TSWG) coordinates the federal research programs designed to help responders, and the Interagency Board for Equipment Standardization and Interoperability (IAB) is in the process of developing agreed-upon standards for emergency responders.[13] As the DHS continues to mature, the development of a more formal RDT&E system for emergency responders will be required.

Within the DHS, the Undersecretary for Science and Technology (US&T) has budget authority, but exact procedures for coordinating federal research efforts by the US&T are still being developed.[14] One aspect of particular concern to emergency responders is the testing and certification of new technologies and equipment.

The Army possesses many and varied testing facilities that, with proper

---

[12]The Army laboratories perform a broad spectrum of research and development (R&D) activities, including basic and applied research as well as advanced development. Testing is conducted by the laboratories as needed, sometimes in their own facilities, and sometimes in other military facilities or in collaboration with other entities such as universities, industry, and other government laboratories.

[13]Additional information about the TSWG is available online at <www.tswg.gov/tswg/about/about.htm and about the IAB at www.iab.gov/page_manager.asp.>

[14]In the DOD, the term "S&T" refers to a program. It is used in that sense in this report. In the DOD, S&T is understood to be the longer-range part of R&D. It consists of three parts, each of which is identified by its own budget category, as follows: (1) basic research, budget category 6.1; (2) applied research, budget category 6.2; and (3) advanced development, budget category 6.3. Together these categories make up the DOD S&T program. Thus, "S&T" in DOD parlance does not include shorter-range R&D in higher budget categories, from 6.4 on up. By contrast, in the DHS the S&T directorate, a distinct part of the DHS established and so identified in the legislation that created the department, funds both short-term and long-term R&D. In the DHS S&T is thus more akin in character and content to what the DOD calls R&D than to what the DOD defines as S&T.

coordination, might be made available to other agencies. For example, the Army has many long-term cooperative arrangements such as that with the permanent groups of engineers at NASA (National Aeronautics and Space Administration) Langley and NASA Glenn using NASA's investment in wind tunnels and other airframe and propulsion testing facilities. If procedures for sharing these facilities with the DHS were put in place, it would greatly assist emergency responders and as a secondary benefit would allow the Army to preserve some underutilized facilities. The committee believes that there are considerable opportunities for collaboration between the Army and the DHS to provide emergency responders with enhanced capabilities and new technology, given the right policy guidance.

SCENARIOS

Scenarios are invaluable tools in helping to determine what capabilities might be required and what types of equipment might satisfy particular requirements. At the present time, national-level scenarios for use in such efforts are not available from the DHS.

To better understand the technology needs of emergency responders, the committee developed four example scenarios that could provide a "mark on the wall" against which to measure C4ISR needs. It emphasizes that these scenarios are intentionally very general, lack significant detail, and should not be interpreted as approved Army, DOD, or DHS scenarios. They are intended merely to illustrate the range of situations against which potential C4ISR needs for emergency responders might be identified. The four scenarios are described below in order of increasing complexity.

Scenario 1: Single Event, Single Location

The first scenario could involve a single event occurring at a specific time at a single location. Examples are the 1995 bombing of the Alfred P. Murrah Federal Building in Oklahoma City or the World Trade Center bombing in 1993. This scenario might also include natural disasters such as the San Francisco earthquake of 1989 or man-made disasters such as the overturning of a truck containing 40,000 pounds of explosive fireworks powder on the National Capitol Region Beltway at the Springfield interchange in Northern Virginia in 1999. The latter incident required closing the intersection of Interstates 95, 395, and 495 for a whole day, resulting in the delay of hundreds of thousands of commuters and the evacuation of nearby residents. With each event such as the truck accident, the potential exists for significant destruction, loss of life, and/or disruption of day-to-day activities, although the event is localized with no follow-on attack or natural disaster.

Scenario 2: Multiple Events, Single Location

The second scenario assumes multiple events occurring over a period of time at a single location. This type of disaster could take the form of an initial terrorist attack followed by terrorist attacks on responders. It could also be an initial terrorist attack followed by some other disaster—for example, the 2001 terrorist attack on the World Trade Center followed by the collapse of the Twin Towers. An example of a natural disaster in this scenario would be the San Francisco earthquake of 1906 that first devastated much of the city with seismic shocks, and then triggered fires that could not be controlled because of building rubble blocking the streets and broken water mains.

Scenario 3: Single Type of Event, Multiple Locations

The third scenario could involve a single type of event occurring over a period of time over multiple locations. Such a disaster could take the form of a natural catastrophe such as Hurricane Isabel, which hit several states on the East Coast in 2003. Or it could include events such as the San Diego wildfires in 2003 that apparently began as a small fire and then, fueled by dry scrub and timber and spread by high Santa Ana winds, developed into multiple, devastating firestorms. These firestorms forced the evacuation of 40,000 residents and required 10,000 firefighters. An example involving a terrorist attack could be the use of a biological weapon at a convention that creates casualties in multiple cities days later.

Scenario 4: Multiple Events, Multiple Locations

The fourth scenario might involve coordinated multiple attacks at multiple locations. An example would be a disaster such as the attacks of September 11, 2001, but with five planes on both coasts. Such a disaster could also take the form of multiple tanker trucks with radioactive debris simultaneously exploding in several places around the nation, destroying government buildings, tunnels, bridges, other infrastructure, and so on. Additionally, the number of locations could increase, as changing winds might carry airborne radioactive debris to other sites.

RELATIONSHIP TO C4ISR CAPABILITIES

In Report 1, the committee found that C4ISR is an important capability cutting across all scenarios. Regardless of the nature or motivation of an attack, it is crucial that command and control, communications, and all aspects of data gathering and analysis be thoroughly coordinated and effective if disaster sites are to be managed properly. In short, emergency responders, just like the military, must *see*, *understand*, and *act*.

In general, the C4ISR requirements for dealing with all disasters are similar, with exceptions for the scope of different events and allowances for the likely incubation periods that may occur in a biological attack. The larger the scope of an event the more complex the C4ISR requirements will be. These needs are difficult enough with a local event handled by local authorities (that is, coordinated among the police, fire, and medical personnel of a single jurisdiction). However, an event drawing in responders from surrounding areas makes C4ISR even more problematic as different techniques, policies, and operational vocabularies are encountered. When higher levels of government such as the military are assisting, the potential for problems is even greater. In Report 1, the committee found such a situation to be very much like the situation in coalition warfighting when the military had to communicate with different forces.

The role of computers in disaster scenarios is ubiquitous. Communications are increasingly computer-based. Command and control will rely on computer-stored data and computer-generated situational displays. Computers play a dominant role in data management. Responders should have detailed building plans and information about the activities conducted in buildings, maps of underground utilities, and locations of shut-off valves and switches available to them in real time. Computers will be necessary in disasters for the surveillance of hospital admissions and other medical records to detect any outbreak of unusual disease patterns that may signify a terrorist attack.

The ISR portion of C4ISR is all of those activities that collect and analyze information about an incident and present ingredients for a common operating picture to decision makers. For emergency responders, the first indication of a serious problem may come from a sensor on a patrol car or from a network of sensors on the city streets or in subway tunnels. The use of multiple sensors and fusion of the sensor data can alert the various responders to the details of the incident. The challenge is to fuse this information into a common operational picture that policy makers can act on. Once the event scene has been established, the incident commander will need as complete a picture of the event as possible. As the crisis management progresses, knowledge of the position, physical condition, and actions of individual emergency responders will be necessary in order to aid in the command-and-control process and to keep the emergency responders away from particularly hazardous locations.

Given technology's increasing capability to provide the incident commander with ever-growing volumes of data, the problems resulting from information overload cannot be overstated. Equally there is the danger that critical information may not reach the proper decision levels where it can be acted on in a timely way. Worse, both phenomena, overload and failure to receive critical items of information, may occur simultaneously. This challenge applies to both Army decision makers and incident commanders and represents another area for collaboration between the Army and civilian emergency responders.

The reasons for seeking new technologies transcend the obvious desire simply to give emergency responders better equipment. The technologies discussed in Chapter 4 of this report allow entirely new capabilities to emerge, such as real-time decision-making ability by the on-scene command team at the crisis site. Such capabilities will enable and strengthen multidisciplinary efforts between and among the various emergency responder groups working on crisis management. These motivations are the very same ones that underlie the Army's transformation program for the Future Force.

Finding 1-2. The Army has developed a number of capabilities that could be used by emergency responders:

- Relevant technologies from the Army science and technology base;
- C4ISR systems that have been developed and deployed by the Army;
- An acquisition system, similar to the Army's spiral development process, that encompasses identifying needs, funding the required technology, and developing fieldable products;
- A testing and certification process for new equipment;
- Training programs;
- A network-centric operations approach;
- Exercises (and supporting facilities);
- Modeling and simulation capabilities; and
- A process for the development and assessment of doctrine.

REFERENCES

CFR (Council on Foreign Relations). 2003. Emergency Responders: Drastically Underfunded, Dangerously Unprepared. New York, N.Y.: Council on Foreign Relations.

Cohen, R. 2003. DOD Homeland Defense. Available online at <http://proceedings.ndia.org/3500/Cohen_Homeland.pdf>. Accessed April 1, 2004.

DHS (Department of Homeland Security). 2003. National Incident Management System, Initial System. July 18. Washington, D.C.: Department of Homeland Security.

FEMA (Federal Emergency Management Agency). 1997. State Capability Assessment for Readiness. December 10. Washington, D.C.: Federal Emergency Management Agency.

Jackson, B., D.J. Peterson, J. Bartis, T. LaTourrette, I. Brahmakulam, A. Houser, and J. Sollinger. 2002. Protecting Emergency Responders: Lessons Learned from Terrorist Attacks. Available online at <http://www.rand.org/publications/CF/CF176/>. Accessed September 24, 2003.

Lippy, B., and K. Murray. 2002. Improving the Training of Skilled Support Personnel for Responding to Terrorist Actions: A Review of the Problems and Feasible Solutions. December 14. Washington, D.C.: National Clearinghouse for Worker Safety and Health Training.

Matthews, K., and J. Buzby. 2001. Dissecting the challenges of mad cow and foot-and-mouth disease. Agricultural Outlook, AGO-283: 4–6.

NORTHCOM (Northern Command). 2003a. Who We Are—Mission. Available online at <http://www.northcom.mil/index.cfm?fuseaction=s.who_mission>. Accessed March 6, 2004.

NORTHCOM. 2003b. Who We Are—Our Team. Available online at <http://www.northcom.mil/index.cfm?fuseaction=s.who_team>. Accessed November 17, 2003.

NRC (National Research Council). 2003. Science and Technology for Army Homeland Security: Report 1. Washington, D.C.: The National Academies Press.

U.S. Army. 2003. Organization: Army Mission. Available online at <http://www.army.mil/organization/>. Accessed November 17, 2003.

2

Capabilities for the Army's Future Force

This chapter describes the Army's Future Force, including the Future Combat Systems Program and the Future Force Warrior Program. These two programs are the current focus of the Army's science and technology efforts.

The Army anticipates that the Future Force will be equipped in such a way that it can exploit the benefits of a network-centric warfare (NCW)[1] mode of military operations (U.S. Army, 2003). A brief discussion of the concept and implications of NCW is provided later in the chapter. The command, control, communications, computers, intelligence, surveillance, and reconnaissance (C4ISR) capabilities of the various system components are critical to achieving NCW operations. Appendix E provides specifics on C4ISR capabilities for the Future Force.

WHAT IS THE FUTURE FORCE?

The Army's Future Force concept is the strategy intended to transform the Army's forces, beginning with platforms and weapons and extending from senior commanders down to individual soldiers. The term "transformational" is used

[1]The committee has employed the term "network-centric" in two separate but related contexts: namely, (1) in warfare, and (2) in homeland security. Thus, the former (network-centric warfare, NCW) is used in Chapter 2, while the latter (network-centric operations, NCO) is used in Chapter 3 and elsewhere when it concerns homeland security. Suffice it to say that in both contexts the term "network-centric" refers to making possible cooperative actions by exploiting latent resources made available only when all participants share information gathered by any one of them. The main goal is to use C4ISR to achieve connectivity to accomplish the mission.

intentionally by the Army in this context to describe innovation on a grand scale, undertaken to address major changes in the character of conflict, to exploit new technologies—particularly information technology (IT) emerging from the commercial world—and to adapt to shifts in geostrategic competition.[2]

For the past 20 years, the Army has been engaged in a transition process to leverage increasingly available innovative technologies, particularly in the area of C4ISR. The strategy calls for developing new kinds of units, modeled on an "Objective Force" that would operate and be organized and equipped differently from today's combat commands (U.S. Army, 2002). In August 2003, the Army Chief of Staff, General Peter J. Schoomaker, redesignated the Objective Force concept as the "Future Force." Army plans for developing the Future Force include a range of leader development, acquisition, training, sustainment, and institutional initiatives. The Army's goal is to field the first fully operational Future Force unit in 2009 (U.S. Army, 2003). Several relevant Future Force C4ISR technologies are already available or are well along in the developmental process. Table 2-1 describes the operational benefits of the Future Force.

CAPABILITIES ENVISIONED FOR THE FUTURE FORCE

The primary capabilities required for the Future Force overall are these:[3]

- Trained soldiers and leaders who understand how to use the power of information and the network to maximize combat effectiveness;
- Situational awareness of all forces—blue (friendly), red (enemy), joint, and neutral;
- A "smart knowledge management system" that knows the user, what the user does, and what he or she needs and that pushes knowledge to the user as well as pulling it from the network when needed;
- Ubiquitous assured access to the network and sensors; and
- The ability to remain relevant to national defense with the maturation of technologies in the commercial world—timely spiral technology insertions.

In broad terms, the Future Force units of action (brigade-size and smaller) will possess the characteristics of responsiveness, deployability, agility, versatility, lethality, survivability, and sustainability. Each of the characteristics described in the following subsections is critically dependent on C4ISR capabilities (U.S.

---

[2]There is no single commonly accepted metric or framework that distinguishes among concepts that are transformational and those that are not. See Roxbourough (2002).

[3]LTG John M. Riggs, Director, Future Force Task Force, "Transformation to the Objective Force: C4ISR Enablers for Homeland Security," briefing to the committee, Washington, D.C., August 25, 2003.

TABLE 2-1 Expected Operational Benefits of the Army's Future Force Concept for the Conduct of Joint Operations

From Past Capabilities	To Projected Capabilities
Stove-piped, staff-centric command-and-control.	Joint-integrated, network-centric battle command; enables decision superiority and self-synchronization.
Fight after force buildup at major air/seaports. Time-consuming force projection.	Immediate employment of forces arriving rapidly through multiple austere entry points.
Sequential, contiguous, linear operations.	Simultaneous operations, distributed throughout joint operations area, within a nonlinear framework.
Attrition-based campaign with massed formations.	Direct attack of centers of gravity with precision effects; defeat through disintegration.
Gaps in situational understanding; uncertainty; intelligence by contact and direct observation.	Global, robust, near-real-time joint intelligence; sensor networks integrated from space-to-mud; improved situational understanding.
Large logistics structure with large forward footprint.	Reduced logistics structure and small footprint through reach-back and distribution-based sustainment.
Effective combined arms operations.	Greater synergy of integrated joint operations.

SOURCE: U.S. Army (2003).

Army, 2001a, 2001b). The characteristics of the Army's Future Force concepts are summarized in Figure 2-1.

Responsiveness

Commanders and staffs will team collaboratively (and virtually) with other elements through the Global Battle Command Network. The development of accurate situational awareness and of a complete understanding of the operational situation and mission begins well before departure from the home station. Awareness and understanding continue to develop while en route to and throughout operations, with updates and adaptations as the situation evolves.

Deployability

The unit of action will be capable of quickly and rapidly concentrating combat power and conducting distributed and continuous combined arms and

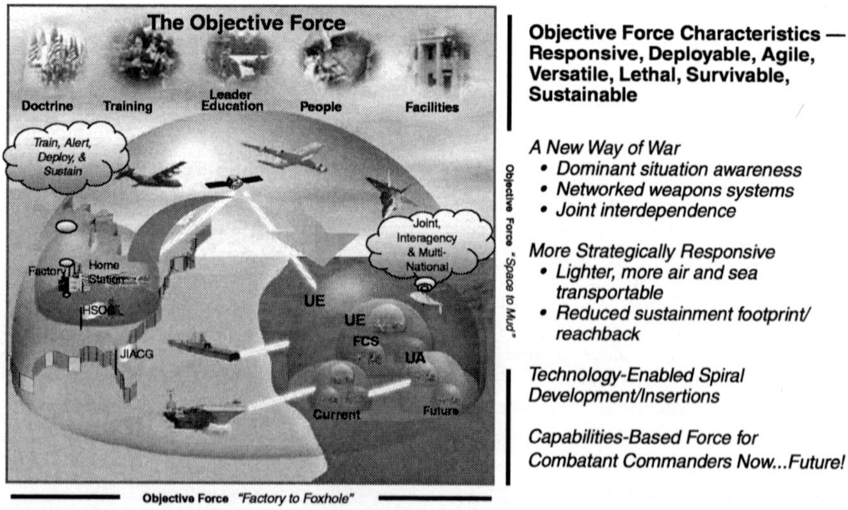

FIGURE 2-1 Characteristics of the Army's Future Force. Courtesy of LTG John M. Riggs, Director, Future Force Task Force, U.S. Army.

full-spectrum operations upon arrival in the area of operations. Prior to departure, training and mission rehearsal can be conducted on the platforms, with embedded virtual training designed to develop individual, crew, and small-unit functional capabilities. Refinements based on intelligence updates can be made while en route.

Agility

The unit of action must be able to make quick transitions to accommodate changes in its task, purpose, and mission. It must be able to maneuver into and out of contact with enemy forces without losing operational momentum. Agility applies to both the mental and the physical qualities needed to meet rapidly evolving battlefield situations.

Versatility

Units must be able to generate formations that can achieve sustained land dominance at any point in the spectrum of warfare, from low-level conflict to full-theater operations. They must be able to do so in all environments, in any kind of weather and by day or night.

Lethality

Lethality requires a triad of sensor effects, force capabilities, and battle command that enables the dynamic application of lethal and nonlethal destructive and suppressive effects to achieve the commander's intent. Networked firepower is fully integrated from theater to platform, with dynamical rerouting of targeting data and missions via flexible, sensor-to-shooter linkages.

Survivability

The unit of action provides maximum possible protection to mounted and dismounted soldiers. Tactics and operations combined with passive and active survival capabilities enable platforms and soldiers to detect and identify potential targets at a distance before being detected themselves, to achieve a kill with the first round fired, and to survive enemy fire if detected and fired upon.

Sustainability

Sustainability requirements will entail the continuous, uninterrupted provision of logistical support to Army forces. This support will be capable of "just in time" rather than "just in case" sustainment, allowing commanders to reduce stockpiles in theater while relying on technology to provide sustained support and real-time tracking of supplies and equipment. Embedded sensors on each platform provide an accurate picture of the sustainment status of the vehicle, weapons systems, and soldier support systems.

NETWORK-CENTRIC WARFARE AND THE FUTURE FORCE

The basic characteristics and capabilities of the Future Force are founded on the concept of network-centric warfare. This concept emerged from the notion of the transition of warfare from the 20th-century Industrial Age to the 21st-century Information Age, as discussed in *War and Anti-War* (Toffler and Toffler, 1993). NCW is a logical extension of previous Army efforts in areas such as the All Source Analysis Center enhanced by the rapid advancement of information technology particularly in the area of C4ISR. The Army intends to use C4ISR technology advancements to connect all weapons systems and sensors and to give U.S. soldiers and commanders the advantage of being able to follow the principles "*See* first, *Understand* first, *Act* first, and *Finish* decisively" during operations.

NCW represents a major conceptual transformation from the traditional Industrial Age approach to warfare, commonly referred to as platform-centric. NCW involves a new and entirely different approach to decision making and operations on the battlefield, affecting everyone from the senior commanders to the individual soldiers. NCW has three overarching characteristics:

- The shift in focus from platforms to networks,
- The shift from viewing actors as independent agents to viewing ensembles of continuously adapting "ecosystems," and
- The importance of making strategic choices to adapt or even survive in a changing ecoenvironment.[4]

Finding 2-1. The network-centric concept is the foundation of the Army's Future Force.

THE FUTURE COMBAT SYSTEMS PROGRAM

The Future Combat Systems (FCS) Program is the Army's top-priority science and technology (S&T) effort. The FCS will constitute the core components of the Army's Future Force. It is a multifunctional, multimission, reconfigurable system of systems that networks soldiers with their commanders, as well as with manned and unmanned air and ground vehicles. By integrating mission capabilities, including direct and indirect weapon use, reconnaissance, troop transport, countermobility, nonlethal effects, secure and reliable communications, and joint interoperability, the FCS coupled with the Future Force Warrior Program will enable soldiers to operate as a coordinated part of a distributed, networked force. The FCS will enable soldiers in the Future Force to perform a wide range of military activities and operations, from small-scale contingencies to stability and support operations, to major theater warfare. The basic elements of the FCS are depicted in Figure 2-2.

THE FUTURE FORCE WARRIOR PROGRAM

The Future Force Warrior (FFW) Program complements the FCS while focusing on the soldier as a system. It is intended to employ open architectures and cutting-edge technologies to develop a revolutionary warfighting system integrated with multifunctional sensors, weapons, physiological status monitoring, and embedded training capabilities to support the individual soldier. This FFW system of systems will evolve to form adaptive, distributed sensor networks for warfighter situational awareness. The soldier systems and subsystems will be integrated into a comprehensive modular fighting package that can be tailored to different mission profiles. Box 2-1 lists the elements of the FFW.

---

[4]LTG John M. Riggs, Director, Future Force Task Force, "Transformation to the Objective Force: C4ISR Enablers for Homeland Security," briefing to the committee, Washington, D.C., August 25, 2003.

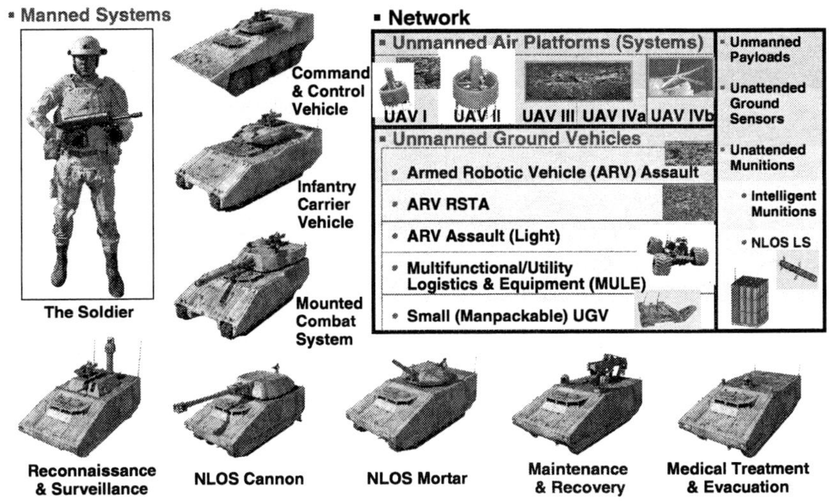

FIGURE 2-2 Basic elements of integrated Future Combat Systems (FCS). Courtesy of LTG John M. Riggs, Director, Future Force Task Force, U.S. Army.

C4ISR CAPABILITIES FOR THE FUTURE FORCE

The C4ISR capabilities required for the FCS are derived from multiple sources (U.S. Army, 2001a, 2001c, 2003).[5] These capabilities (fully portrayed in Appendix E) have been recast using the C4ISR taxonomy of this study: C3 (command, control, and computers) required for timely understanding and informed decision making by commanders and warfighters conducting operations; C (communications) to provide essential networking connectivity to deliver timely information to each warfighting decision maker; and ISR (intelligence, surveillance, and reconnaissance) to supply the platforms, systems, and sensors that collect, fuse, analyze, and interpret the battlefield situation.

Conclusion 2-1. The U.S. Army possesses a large and varied number of Future Force science and technology programs that, with proper coordination, could be made available to the Department of Homeland Security; however, there is currently no planning process to identify which could be shared or how to do so.

[5]LTG John M. Riggs, Director, Future Force Task Force, "Transformation to the Objective Force: C4ISR Enablers for Homeland Security," briefing to the committee, Washington, D.C., August 25, 2003.

BOX 2-1
Future Force Warrior Elements

Lethality. Direct and indirect engagement; less than lethal engagement; target detection and recognition; synchronization of fires; target handoff; ID friendly/enemy/noncombatant; target designation.

C4I. Situational understanding; information management; communications; enhanced vision and senses; detect and avoid hazardous areas; area denial; mark items of interest; intelligence collection and dissemination; mission planning and rehearsal.

Power Sources. High-density, lightweight, efficient, safe, reliable power (includes hybrids and rechargeables).

Analysis and Assessment. Modeling tools to enable optimal system development and assessment; virtual prototyping; individual and force on force modeling.

System Engineering and Integration. Integrate all areas into comprehensive, integrated system of systems. Weight, power, and cost treated as independent variables.

Survivability. Full spectrum individual protection; signature management; thermal management; physiological status monitoring.

Mobility. Horizontal, vertical mobility; reduce and offload equipment carriage; identify, reduce, and defeat obstacles; position/location/tracking.

Sustainability. Delivery of tactical resupply; water purification and generation; water management.

Training. Individual, small unit, leader training concepts; embedded training, novel training, tactics, and procedures to exploit Future Force Warrior capabilities.

Human Performance. Sustain and enhance individual and team performance; optimize system and team fightability; optimize human endurance, cognitive, and physical capabilities.

SOURCE: Adapted from Carol Fitzgerald. 2003. Future Force Warrior. Presentation by Program Manager for Future Force Warrior Technology at the Future Force Warrior Fightability Workshop, Framingham, Mass., August 18.

Recommendation 2-1. The U.S. Army, through the Department of Defense, should work with the Department of Homeland Security to analyze and determine, among other items, appropriate planning processes necessary to determine which Future Force science and technology programs should be shared and how best to go about doing this.

SUMMARY

To "*See* first, *Understand* first, *Act* first, and *Finish* decisively" in support of joint and combined operations requires a multifunctional, multimission, reconfigurable system of systems that networks soldiers with their commanders as well as with manned and unmanned air and ground vehicles. The underlying theme of the Army's transformation is the focus on marrying capabilities driven by new technologies, including new ideas in economics, information technologies, and business practices, to the evolving systems. For maximum impact, rapid decision making requires attention to responsiveness, deployability, agility, versatility, lethality, survivability, and sustainability.

Many of the same capabilities that contribute to network-centric warfare for the Army may be adaptable as capabilities for homeland security. The capabilities needed for emergency responders are the subject of the next chapter.

REFERENCES

Roxbourough, I. 2002. From Revolution to Transformation: The State of the Field. Available online at <http://www.dtic.mil/doctrine/jel/jfq_pubs/1332.pdf>. Accessed October 2, 2003.

Toffler, A., and H. Toffler. 1993. War and Anti-War: Survival at the Dawn of the 21st Century. Boston, Mass.: Little, Brown.

U.S. Army. 2001a. Statement of Required Capabilities for Future Combat System of Systems (FCS). November 2. Fort Monroe, Va.: Training and Doctrine Command.

U.S. Army. 2001b. The United States Army Future Force Tactical Operational and Organizational Concept for Maneuver Units of Action, Draft version 2. TRADOC Pamphlet 525-3-91. November 6. Fort Monroe, Va.: Training and Doctrine Command.

U.S. Army. 2001c. Mission Need Statement for Future Combat System of Systems (FCS) Potential ACAT 1. November 2. Fort Monroe, Va.: Training and Doctrine Command.

U.S. Army. 2002. Military Operations: Future Force Maneuver Units of Action. TRADOC Pamphlet 525-3-90. November 1. Fort Monroe, Va.: Training and Doctrine Command.

U.S. Army. 2003. The Army Future Force: Decisive 21st Century Landpower. August. Fort Monroe, Va.: Training and Doctrine Command.

3

Capabilities for Emergency Responders

The purpose of this chapter is to identify the command, control, communications, computers, intelligence, surveillance, and reconnaissance (C4ISR) requirements for emergency responders. It addresses currently lacking capabilities as well as emerging future requirements. This chapter describes the scope of the emergency responder community considered, the tasks that this community could be required to perform, the conditions under which these activities might occur, the characteristics and functionality of the C4ISR technologies that responders would need in order to deal with the consequences of a disaster or a terrorist incident, and the training and exercise opportunities that currently exist. Lastly, it describes Project Responder, an independent effort focusing on the status of equipment for emergency responders. Appendix F provides specifics on C4ISR capabilities needed by civilian emergency responders.

ABILITY TO RESPOND TO MANY THREATS

The committee examined the requirements of emergency responders—that is, the personnel and services constituting the national response capabilities that could be called on to deal with a disaster or a terrorist attack. The committee acknowledges that this evaluation was accomplished at a fairly high level of abstraction. It is possible that different conclusions might be drawn should a highly detailed examination be conducted. A term in common usage, "first responders," usually refers to law enforcement, firefighting, and emergency medical personnel.[1] These responders, however, are not the only assets that may be required in the aftermath of an attack on the homeland. In contrast, the term "emergency responder" encompasses all personnel within a community who could be needed in the event of a natural or man-made disaster or a terrorist incident (LaTourrette et al., 2003). As indicated in Chapter 1, the Homeland Security Act of 2002 defines emergency response providers as including "federal, state, and local public safety, law enforcement, emergency response, emergency medical (including hospital emergency facilities), and related personnel, agencies, and authorities" (P.L. 107-296, Sec. 2(6)).

In addition, commercial assets such as communications industries and private, nonprofit, nongovernmental organizations (NGOs) such as the Red Cross or the Salvation Army can also play an important role in emergency response. However, there is no national effort to leverage the supporting technological capabilities of these organizations for an effective response to a disaster or a terrorist attack.

While it is believed that about 2.3 million firefighters, police, and emergency medical personnel could be considered emergency responders, these numbers do not suggest the full scope of the national response force (LaTourrette et al., 2003). Some have estimated that the broader public emergency response community could be as numerous as 9 million to 10 million.[2] In addition to professional responders and volunteers, there is, for example, a pool of about 6.5 million skilled construction workers in the United States who could potentially be called up to respond in the wake of disasters. All of these assets could benefit from the enhanced use of C4ISR technologies. The sheer number of responders speaks to the immediate need for a compatible C4ISR architecture and standards set to coordinate and prioritize the activities of multiple response entities. Consideration should be given to identifying a simple but executable and expandable architecture as a start. The vast number of responders also suggests that significant economies of scale could possibly be achieved in terms of reducing unit costs for purchasing and maintaining emergency responder support systems and equipment.

> **Conclusion 3-1.** Once fully established, the national requirements for command, control, communications, computers, intelligence, surveillance, and reconnaissance (C4ISR) technologies to support emergency responders will be substantial and sustainable and could create a significant market.

---

[1]See, for example, U.S. Congress (2002). There is, however, no common definition of "first responder." For example, the "National Strategy for Homeland Security" refers to first responders as police, firefighters, emergency medical providers, public works personnel, and emergency management officials. See OHS (2002).

[2]See, for example, Collins (2000).

Communications equipment and capabilities vary widely. The larger cities of the nation appear to be reasonably well equipped for meeting disasters, but more rural and smaller jurisdictions receive little or no help in obtaining what they may need. Washington, D.C., for example, is perceived to be a high-value target and therefore receives much support. The nation's capital has a sophisticated communications capability, including two-way pagers for senior leadership, appropriately scrubbed intelligence information provided to the responders who need it, satellite phones, an 800-megahertz (MHz) communications system, and full broadband multimedia capability. The city also has two Emergency Operations Centers and several mobile command posts, the newest with full multimedia broadband capability.[3] This is not the case in rural and small jurisdictions. Federal grants are passed through states, and in the past some of those resources have been "skimmed off" to meet valid state requirements. Most recent federal grants have been given with restrictions concerning the amount that states can retain (e.g., 20 percent). Local municipalities generally have a list of shortages in needed equipment and capabilities by functional area but have often done little or nothing to prioritize the list on the basis of a multifunctional, all-hazards approach to mitigating damage. The net result is an uneven national approach to the funding and fielding of the technologies that would be needed if there were a prioritized approach to a common operational framework.

Considering the scope of the emergency responder requirements, there may be significant advantages to be gained by employing C4ISR technologies that would link responders into a system of systems similar to the Army's vision for linking the capabilities of its ground forces and integrating them with the capabilities of the other military services and coalition partners. As suggested in Chapter 2, the military's network-centric approach to operations could serve emergency responders equally well. Such a system could produce significant efficiencies in terms of shared skills, knowledge, and scarce, high-value assets. Such an approach would build capacity and redundancy in the national emergency response system as well as gaining the synergy of providing a common operating picture to all responders and allowing them to share information readily. Network-centric systems could be particularly valuable for responding to large-scale attacks or those involving multiple weapons of mass destruction (WMD). In such situations, responders would have to surge capacity quickly, adapt to difficult and chaotic conditions, and respond to unforeseen requirements.[4]

In short, the committee believes that emergency responder needs suggest that the national emergency response system develop and adopt a network-centric

---

[3]Michael Sellitto, Deputy Chief for Special Operations, Washington, D.C., "C4ISR for the Washington DC Fire Department," briefing to the committee, Washington D.C., July 21, 2003.

[4]For the scope of assets that could be required to respond to a WMD incident, see Larson and Peters (2001).

operations (NCO) approach. The committee defines "network-centric operations" as an information-enabled concept of operations that generates increased operational effectiveness by networking sensors, decision makers, and emergency responders to accurately see, understand, and act on the situations facing them. In essence, NCO translates information superiority into operational power, effectively linking knowledgeable entities in the response to emergencies from the local to the national level.

Conclusion 3-2. Individual emergency responder C4ISR systems need to be linked and integrated into a national operational framework.

Recommendation 3-2. The U.S. Army, through the Department of Defense, should offer to assist the Department of Homeland Security in developing a concept of operations for a national operational framework, to include the appropriate architectures and enabling technologies for C4ISR.

ABILITY TO CARRY OUT A WIDE RANGE OF TASKS

A wide range of emergency responder tasks could be facilitated by C4ISR technologies. The National Strategy for Homeland Security defines six critical mission areas: (1) intelligence and early warning, (2) border and transportation security, (3) domestic counterterrorism, (4) protecting critical infrastructure and key assets, (5) defending against catastrophic threats (i.e., research and development for the other five critical mission areas), and (6) emergency preparedness and response. This report is focused on C4ISR needs in support of the sixth function, emergency preparedness and response,[5] which includes the preparation for, response to, and recovery from a disaster or terrorist attack. The assessment presented in this chapter includes C4ISR needs for planning, logistical support, maintenance and diagnostics, training, and management, as well as C4ISR needs for supporting the actual activities at a disaster site and for addressing post-recovery lessons learned. It should also be emphasized that this study considered C4ISR support for all emergency response functions that take place during an incident, not just that of setting up integrated command and control for incident commanders. In particular, the committee's assessment found that there could well be significant intelligence, surveillance, and reconnaissance (ISR) functions associated with a response, creating a common operating picture to help responders avoid threats or ensuring that they are equipped "just in time" to address threats.

---

[5]The first report in this series (NRC, 2003) talked about "recovery and consequence management technologies"; emergency preparedness and response terminology is adopted in this second report to conform to the National Strategy for Homeland Security (OHS, 2002) and to highlight the important pre- and post-event requirements of emergency response.

Emergency Preparedness and Response Tasks

The following list of emergency preparedness and response tasks is adapted and modified from the list in Report 1 (NRC, 2003, pp. 93-94). The tasks outlined below are generic. They are not intended to refer to a particular type of emergency responder or level of response. In addition, the tasks outlined here may not be accomplished in distinct phases or may be limited to only one phase of an emergency operation.

Throughout Event

- Gather information; and
- Provide continuous public information.

Pre-Response

- Evaluate lessons learned from previous incidents;
- Conduct vulnerability and risk assessments of response activities and response support infrastructure;
- Plan a response;
- Establish communications protocols;
- Train for disaster or terrorist attack response;
- Coordinate with other agencies, levels of government, and private sector assets;
- Establish procedures, including the use of sensors and other means to monitor critical support infrastructure, as required;
- Maintain information on critical infrastructure and geospatial data on areas of interest;
- Provide acquisition and logistical support;
- Perform maintenance, testing, and diagnostics; and
- Provide continuous public information.

Initial Response

- Deploy responders;
- Protect responders;
- Establish an information clearinghouse;
- Monitor location and status of responders;
- Identify the incident commander;
- Establish an interoperable C4ISR system with existing assets;
- Assess in real time the extent of the physical damage, casualties, and the enduring level of contamination and risk of disease transmission;
- Establish quarantine zones, safe areas, and perimeter control of movements;

- Triage and treat the injured;
- Conduct crime scene management; and
- Provide continuous public information.

Containment

- Expand area of control and model and/or predict hazardous areas;
- Isolate secondary threats (ruptured gas mains, interrupted electrical service, instability of damaged infrastructures and buildings);
- Restore or replace infrastructure critical to containment;
- Restore and maintain C4ISR systems with restored or replaced infrastructure;
- Perform environmental monitoring;
- Conduct a site survey, determine additional needs, and provide reinforcements;
- Provide continuous public information; and
- Maintain C4ISR interoperability.

Near-Term Recovery

- Provide continuous public information;
- Eliminate and/or control the ongoing immediate threat (e.g., contain the effects of weapons of mass destruction);
- Expand the treatment of casualties (begin stress management, including that for responders) and evacuate the injured;
- Rescue, protect, evacuate, and track civilians;
- Manage the identification, tracking, and reunification of missing persons;
- Conduct mortuary operations;
- Assure food and water safety;
- Provide food, shelter, and support for personnel in the affected area;
- Determine, marshal, and deploy assets required for long-term operations;
- Conduct additional training for emergency responders for site-specific threats;
- Manage volunteer resources;
- Provide additional geospatial resources;
- Provide emergency veterinary services and support for animal and plant control and disposal; and
- Establish a sustainment base.

Post-Event Recovery

- Disengage responder assets;
- Consolidate and redeploy assets;

- Provide maintenance and logistical support;
- Conduct an after-action study and maintain a record of lessons learned;
- Reconstitute assets;
- Update plans;
- Retrain assets; and
- Identify new organizational or material requirements.

Restoration of Normalcy

- Provide decontamination support;
- Provide financial management for responder resources and manage contractual support;
- Provide post-event counseling;
- Restore public order and essential services;
- Assess casualties, damage, and environmental impact;
- Treat mass casualties;
- Restore the physical infrastructure; and
- Provide continuous public information.

ABILITY TO FUNCTION EFFECTIVELY IN A DANGEROUS AND/OR CHAOTIC ENVIRONMENT

In determining the technological needs of emergency responders, the conditions under which operations occur must also be considered. For example, emergency response operations may be conducted under the same chaotic conditions characteristic of a battlefield, particularly in an urban environment. As with combat forces, emergency responders in crisis situations may find it difficult to communicate with and determine the location of their organizations. Many current communication and locator systems, for example, are dependent on line-of-sight technologies that are easily disrupted by tall buildings and underground infrastructure. In addition, in an urban environment both responder and combat forces face the challenges of a limited area of observation, restricted span of control, and canalized movement. These restrictions have significant implications for the speed, size, and efficiency of operations. Complex terrain (buildings, elevated highways, and so on) as well as the physical destruction resulting from combat or a disaster or terrorist attack will force dispersion of forces or responders, nonlinear operations, and decentralized control, limiting the ability of assets to coordinate, reinforce, or support one another.

In fact, the emergency responder environment suggests that advanced C4ISR capabilities prized on the battlefield could also be essential to improving national emergency response capabilities. In addition to the direct benefits of C4ISR, these capabilities can have significant indirect benefits for other aspects of operations. For example, on the battlefield, the knowledge gained from advanced

C4ISR can be used to reduce materiel requirements. Soldiers use battlefield knowledge to avoid threats and decrease requirements for munitions and armor protection, and responders can use information systems to reduce needs and improve on the capabilities of personal protective equipment. One relevant concept is "just in time" logistics—that is, the ability to ensure that support arrives at the scene precisely when it is needed rather than having resources stockpiled or requiring responders to carry equipment with them all the time.

Another feature common to battlefield and emergency responder environments is the utility of situational awareness. The military expects that maintaining a common operational picture will allow its troops to avoid threats, and emergency responders may likewise rely on early warning to minimize their exposure to risks and decrease requirements for personal protective equipment and other support assets. Additionally, emergency responders could benefit from C4ISR capabilities similar to those needed by high-tech warriors: for example, reduced weight and power-generation requirements, non-line-of-sight systems, hands-free controls, and heads-up displays. Finally, C4ISR capabilities that are backward-compatible to older systems and technologies are essential to ensure the viability of the high-low technology mix.

C4ISR CAPABILITIES FOR EMERGENCY RESPONDERS

Given the personnel, tasks, and conditions outlined above, the committee identified shortfalls in the capabilities required by emergency responders in the area of C4ISR.[6] This section identifies characteristics and functionalities of C4ISR technologies needed by emergency responders.

Command, Control, and Computer Capabilities

As to specific shortfalls in the areas of command and control, the greatest emergency response needs are in the capacity to scale responses to events that can range from local disasters to terrorist attacks involving catastrophic WMD attacks. Key elements in managing the scope of the response are as follows: to be better prepared before an emergency with better intelligence and training; to be able to assess a situation rapidly; and to be able to share this information with other authorities, which would include being able to hand over control to other authorities, if necessary. Needed capabilities include the following:

- *To be able to see first.* There is need for rapid and accurate situation assessment and the ability to produce a continually updated common operational picture. The common operational picture not only must have

[6]This effort was complicated by the lack of uniform national standards that define the regional, state, and local capacities needed to respond to a terrorist attack. See Canada (2003).

the appropriate information for decision making but also must be presented in a way that highlights the most-time-critical information. The common operational picture should be able to display the nature and number of one's own forces, the risks they face, and the facilities and services, including communications, at their disposal.

- *To be able to understand first.* There is a need for access to information—not only intelligence, but also background information that may be critical to handling the crisis. This could include information about infrastructure, facilities, and resources: for example, knowing the locations of hazardous materials, having floor plans of structures, and being aware of key personnel with critical knowledge. It is necessary to be able to provide information on the location and status of responders within the disaster area and of reinforcing responders from other jurisdictions. Tools for collaboration among responders are needed.
- *To be able to act first.* There is need for decision-making aids that can access, query, evaluate, and make recommendations employing large amounts of information maintained in different databases and transmitted by various communications systems.

Significantly, many elements of the command-and-control programs for managing military operations for the Future Force call for capabilities similar to those listed above. Much as the military envisions using its future command-and-control systems as a linchpin for conducting network-centric warfare, the committee believes that emergency responder command-and-control systems could provide the basis for emergency responders to benefit from the effectiveness of network-centric operations.

Many of the command-and-control capabilities for emergency responders should be based on published standards in order to facilitate broad cooperation and coordination among state, local, and federal response assets as well as with capabilities from the private sector. In addition, the committee concluded that command-and-control systems also require a degree of assurance and redundancy and that they must be resilient against critical infrastructure failures, particularly the loss of access to the Internet and wireless networks.

Additionally, there is an important need to address significant shortfalls in command-and-control functions related to responding to large-scale WMD attacks. A common concern of responders is the need for effective perimeter control at the scene of an event in order to provide for management of movement within the site to facilitate operations and avoid hazards, and the need for control of traffic to accomplish evacuation away from the site. Significant unresolved problems in site management for catastrophic events or terrorist attacks also include those of processing patients, accounting for missing persons, and managing the volunteer support and the housing needs of displaced persons. The need for pre-disaster training, including realistic, high-quality exercises that cover multiple jurisdictions and levels of government (including the employment of defense assets under the control of the U.S. Northern Command) are also cited by responders as an urgent requirement. Finally, command-and-control systems require means of support and sustainment to ensure a high degree of operational readiness. The ability to sustain a robust response to large-scale terrorist attacks will likely depend on logistical capabilities (Jackson et al., 2002).

Computer and software support for emergency response is also inadequate at the present time to deal with large-scale disasters or terrorist strikes. Computer systems for emergency responders are envisioned as providing the incident commanders with an integrated view of information relevant to a disaster scene. Global Information System (GIS) databases are expected to play an important role in presenting a combination of static information (such as building layouts, floor plans, connections to utilities) and dynamic information (such as locations of emergency responders, fire conditions, and so on) (Beakley;[7] Cashin et al., 2003). GIS systems are developed for the overlaying of static information and may be updated weekly or monthly. The DHS's Directorate of Emergency Preparedness and Response (EPR)/Federal Emergency Management Agency (FEMA) has a close working relationship with the National Geospatial-Intelligence Agency for remote sensing/GIS support. The available capabilities include many state-of-the-art and ongoing research and development (R&D) efforts in the remote sensing/GIS arena. However, they are not currently at the point of overlaying dynamic information that is continually updated as an event unfolds.

Setting the standards for the necessary databases will be an important part of developing computer systems used by emergency responders. Another aspect will be the method by which emergency responders and utility workers can update the information in the databases as a result of routine inspections. For example, firefighters make handwritten notes on conditions inside buildings during inspections. These notes are not standardized and may not even be legible to others in times of emergency. Instead, one can imagine the use of a voice-activated personal digital assistant (PDA) to make entries in a standard format that can be downloaded into an appropriate database.

While a wide range of computing hardware is readily available today at affordable prices, the challenge is to integrate the hardware and software into an interoperable system of systems. Particular challenges that emergency responders face include these:

- *Interoperability*—probably the greatest challenge. In this area, following commercial standards makes the most sense. Proprietary protocols and systems should be avoided.

---

[7]Guy Beakley, Hicks and Associates, Inc., "C4ISR Requirements for the Nation's First Responders from Project Responder," briefing to the committee, Washington, D.C., July 22, 2003.

- *Processes for scaling up as the magnitude of a crisis builds.* This requirement relates to the need to know who is in charge and how to transition authority.
- *Exercises to determine whether the systems work as anticipated.*
- *Ability to protect sufficient command-and-control infrastructure and capabilities during emergencies.*

Communications Capabilities

Many reports have cited emergency responders' needs for secure voice, video, and data communications that are interoperable among agencies and governmental affiliations as well as scalable with the size of the event (Cummings;[8] Cashin et al., 2003; NIJ, 2003; Schwabe et al., 2001). At a response scene, police, emergency medical services, and firefighters often use incompatible radio systems (LaTourrette et al., 2003; ISTS, 2001). Because millions of emergency responders are spread over thousands of state and local agencies and departments, they lack the organization necessary to produce a vision of future needs and possibilities (NIJ, 2003, p. 10).

As a result of the heightened interest in homeland security resulting from the tragic events of September 11, emergency responder departments in the nation's largest cities are now developing such a vision (Cashin et al., 2003). The principal strategy for addressing the need for interoperability has been to push for the implementation of a uniform, digital, 800-MHz backbone system. These systems, however, have not proven to be a "silver bullet." While they have many advantages over traditional analog radio systems, concerns include their high costs, their inability to communicate effectively in complex urban terrain, and their inability to prioritize voice traffic (LaTourrette et al., 2003).

Emergency responder communications systems are currently trapped by the history of their development into narrowband channels that are "inadequate and scattered widely in 10 discrete bands across the spectrum, making it difficult for different agencies and jurisdictions to communicate" (NIJ, 2003, p. 10). In any activity, emergency responders need the ability to communicate among themselves in the manner that best serves the functioning of the individual units. In major events it will be necessary for local emergency responders to coordinate with neighboring units, utility workers, state agencies, the National Guard,

---

[8]John C. Cummings, Department of Homeland Security, Science and Technology, "An Overview of the Department of Homeland Security," briefing to the committee, Washington, D.C., August 26, 2003; Guy Beakley, Hicks and Associates, Inc., "C4ISR Requirements for the Nation's First Responders from Project Responder," briefing to the committee, Washington, D.C., July 22, 2003; Donald C. Mertz, Director of Command and Control, Communications, and Computers Systems, Joint Task Force—Civil Support, "Communications Interoperability Between Military and Civilian Agencies," provided to the committee for the August 25-26, 2003, meeting.

FEMA, and others. For example, the members of a small group of firefighters have to communicate among themselves, while their incident commander may need to communicate with the operations center, police, other fire departments, and others.

Current communications systems frequently do not work well in providing assured communications among emergency responders in environments such as tunnels, basements, and high-rise buildings. The installation of infrastructure equipment, such as repeaters, can overcome many of these difficulties. However, dependence on such infrastructure may leave the communications system vulnerable to failure in the event of neglected servicing, power outages, and accidental or malicious damage. Thus, the communications system must have the robustness to withstand the failure of individual nodes and to provide coverage in difficult environments.

Video and data communications have been implemented for limited application by emergency responders, but far greater applications are already envisioned. Some but not all emergency responders have access to video data from helicopters flying over disaster scenes. It is easy to foresee the use of unmanned vehicles, both terrestrial and airborne, to provide incident commanders with video images of disaster scenes from vantage points that would be dangerous or difficult for humans to access. In addition, it may be useful for incident commanders to obtain video images of the conditions faced by emergency responders transmitted from helmet-mounted cameras carried by the responders.

Data communications to police cruisers are now widespread, allowing police to access criminal activity databases in their motor vehicles, and emergency medical service workers now have the capability of sending information about patients needing treatment ahead to hospitals. However, many firefighters lack any data communications capability in their vehicles. For all emergency responders, it is easy to imagine more far-reaching types of data communications (Cashin et al., 2003)—for example, GIS information about roads, buildings, utilities, and other infrastructure needed at a disaster scene.

Data from many databases need to be communicated and presented in a comprehensible way to emergency responders. For example, situational awareness information needs to be transmitted to the incident commander, with reachback to other command centers and unit headquarters.[9] Another example of the need for data communication involves the transmission of information about the health and equipment status of individual emergency responders to the incident commander and unit headquarters and the return to the emergency responder of warning signals, such as evacuation orders.

Communications with sensor networks may also play an important role in the activities of emergency responders. In the case of police, such networks may

[9]Guy Beakley, Hicks and Associates, Inc., "C4ISR Requirements for the Nation's First Responders from Project Responder," briefing to the committee, Washington, D.C., July 22, 2003.

be used to detect unauthorized intrusion into an area. For firefighters, smoke and heat sensors distributed throughout a building may give information on fire conditions. In a major event, emergency medical services may use patient sensors to monitor the locations and conditions of injured people who are awaiting treatment. Technologies for locating emergency responders at the scene of a disaster may also make use of sensor networks.

Sensor networks that are permanently installed, as in a building, may be linked together by cable. However, for economic reasons (e.g., when covering larger areas), it may be desirable to use radio links. For any system, its link to the incident commander may best be made via wireless communications. In the case of intrusion detection and locating of emergency responders, the detection process may involve radio technology at the sensor. Sensor networks that are established at the time of an incident will most likely be linked using radio technology.[10]

During emergency conditions, some emergency responders may be fitted with devices to monitor medical conditions, equipment status, and environmental conditions, to provide warning alarms, and to display location information and evacuation routes.[11] These capabilities will require hands-free voice recognition, noise cancellation, and so on. Providing these functions and connecting the devices to a radio for transmission will require a specially adapted computer, worn by the individual. Most importantly, national emergency responder communications systems require an overarching enterprise architecture. This system must have the resiliency and redundancy to enable continued operations even if the network is directly targeted and attacked by terrorists. It must be capable of establishing priority communications links and addressing the "overload" demands that could occur during a crisis response. It must be open architecture, such as the one proposed by the DOD's C4ISR Advanced Concept Technology Demonstration.

Intelligence, Surveillance, and Reconnaissance Capabilities

While enhanced command, control, computers, and communications are essential to developing the shared common operational picture needed to enhance the capability of federal, state, and local emergency responders across the nation, the capabilities that support this picture are equally essential and indispensable. They are founded on the integration and analysis of the products of multiple ISR sensors and the firsthand reports of emergency responders and observers.

---

[10]A recent issue of *Proceedings of the Institute of Electrical and Electronics Engineers* (Volume 91, Issue 8, August 2003) is devoted entirely to sensor networks and applications. Nearly all of the work reported was supported by the DOD, some of it by the Army.

[11]Guy Beakley, Hicks and Associates, Inc., "C4ISR Requirements for the Nation's First Responders from Project Responder," briefing to the committee, Washington, D.C., July 22, 2003.

The ISR system of systems for homeland security emergency responders can contribute to a shared common picture by helping to meet several basic requirements, including the following:

- Establish databases under normal conditions to serve as a template for comparison,
- Facilitate common situational understanding,
- Monitor critical assets required for response,
- Provide event assessment, and
- Conduct course-of-action development and situation management.

Technologies that help meet these fundamental purposes are essential to developing a truly effective national emergency response capability.

Fielding an ISR family of systems along with the requisite displays and analytical tools may be well beyond the initial capacity of state and local agencies. However, it is possible and desirable to build a national capability based on broadband communications drawing on information from selected assessment centers.

Even before an event, ISR systems should ideally provide the intelligence needed by emergency responders to prepare for operations and the ability to identify the agency and/or the officials responsible for collecting and analyzing different types of intelligence. Intelligence collection will require implementing tools, training, and processes to support intelligence activities beforehand. In addition, well before a crisis C4ISR should provide the capacity for the early detection, identification, assessment, and tracking of, for example, exposure to biological agents through epidemiological and veterinary surveillance.

Determining the extent of the physical damage from an attack or disaster involves comparing the resultant damage to the original status of facilities in the area. Databases that describe the design of facilities and their location are important in establishing the baseline condition. In the future, it is possible that structures will have embedded sensors that measure stresses occurring as the result of both natural and terrorist events. Overhead imagery can be used to systematically describe the effects of an event. Additionally, local terrestrial sensors can be placed in an affected area to provide focused readings of the effects. Unmanned robotic vehicles may enhance the common operational picture of natural or terrorist events. The ability to fuse the measurements from overhead imagery and focused sensor reports from embedded sensors can enable emergency responders to rapidly build an understanding of the magnitude of an event.

One of the most critical tasks following a WMD event would be to assess the extent and spread of the chemical, biological, or nuclear contamination. The development of the courses of action for containment, remediation, and decontamination are highly dependent on the current and projected status of the contamination. The characterization of the contamination is developed from knowledge of the location of the event, environmental factors, and the results of multiple sensor readings. It is plausible that plume models could be employed to provide real-time and projected contamination contours. These contours would then be displayed on digital maps to create a common operational picture of a WMD event. A shared picture of this kind can greatly enhance the process of developing courses of action for containment, remediation, and decontamination.

Unfortunately, there are few sensors distributed around the nation for chemical or biological events (however, a fixed infrastructure of nuclear sensors is distributed across the nation). The U.S. Northern Command (NORTHCOM) would probably be called upon to provide and disperse sensors to characterize a WMD event. Overhead imagery could be useful in the process. It is feasible that in the future a family of unmanned aerial vehicles could be employed to plant a family of sensors in a contaminated area to continually update and assess the situation. And in some cases, emergency personnel in protective suits might insert terrestrial sensors to characterize the situation.

Another aspect of the assessment process is the determination of the status of casualties. In a chemical, biological, radiological, nuclear, or high-explosive event, emergency responders will need to know the status and location of the many potential casualties. Much of this picture will be generated from databases and estimates of similar events, but as happened when the World Trade Center's Twin Towers, parts of the Pentagon, and the Murrah Federal Building were destroyed, there can be an urgent need to locate casualties buried in rubble. Although this situation also occurs with earthquakes, the task of locating and rescuing people is far more complex if an area is contaminated as a result of a WMD event. In the Army's science and technology (S&T) program for the Future Force and for urban combat, new sensors are being developed to "see" inside structures, and robotics equipped with sensors are being developed to go inside structures and under rubble and debris.

The benefits of blue force tracking and in-transit visibility, which allow participants to know where personnel are located, have been clearly demonstrated in recent conflicts. These systems can also contribute to the development of a common operational picture for emergency responders. In planning a course of action for emergency responders, it would be very useful to know where emergency responder vehicles, food stocks, medical supplies, and safe facilities are located and what their status is. Accounting for the location of responders at the incident scene is considered a significant challenge. Many firefighters, for example, are injured and do not receive prompt treatment because of confusion over the location and activities of individuals on the scene. Another concern is the possibility of physical assault. Responders focused on providing aid to victims and managing an on-scene response believe that they could be particularly vulnerable to surprise attacks and other violent acts (LaTourrette et al., 2003).

Fire alarm systems in large buildings offer an example of a limited ISR system that is already used by emergency responders. Improving this system to provide firefighters with information on temperature, smoke, and other aspects of a situation can be financed by building owners as a building code requirement. Easy interpretation of this information would require the additional development of graphical displays and integration with GIS databases. The National Institute of Standards and Technology foresees the integration of fire prediction models with building and fire information to predict the advance of fires as a firefighting tool, much as plume models might be used to deal with WMD events.

OPPORTUNITIES FOR TRAINING AND EXERCISES

Training

Emergency responders receive the majority of their training opportunities in their own communities. However, the DHS, through the Office of Domestic Preparedness (ODP) and the Federal Emergency Management Agency, provides direct training and technical assistance to state and local jurisdictions to enhance their capacity and readiness to respond to domestic incidents as part of the State and Local Domestic Preparedness Training and Technical Assistance Program. Based on National Fire Protection Association standards, the training provides emergency responders with comprehensive instruction in the areas of WMD awareness, technical support, operations, and terrorist incident command.

All courses are reviewed rigorously by federal, state, and local subject matter experts who examine the course materials to ensure their accuracy and compliance with accepted policies and procedures. ODP staff have established regular and recurring meetings with representatives from the Federal Bureau of Investigation, the Centers for Disease Control and Prevention, the Public Health Service/Office of Emergency Preparedness, and the National Fire Academy to discuss and coordinate the development of training for responding to WMD attacks and the delivery of such training courses. Additionally, ODP has on-site representation from the National Guard Bureau to coordinate program efforts and provide technical assistance and guidance.

Of note, the Institute of Medicine's *Preparing for Terrorism—Tools for Evaluating the Metropolitan Medical Response System Program* (IOM, 2002) provides an excellent description of programs of training for medical-emergency first responders, as well as an evaluation of the effectiveness of that training and other elements of preparation in relation to response scenarios. These efforts go well beyond training of individual medical personnel within the framework of their individual responsibilities.

Exercises

Experience and data show that exercises are a practical and efficient way to prepare for crises. They test critical resistance, identify procedural difficulties, and provide a plan for corrective actions to improve crisis and consequence management response capabilities without the penalties that might be incurred in a real crisis. Exercises also provide a unique learning opportunity to synchronize and integrate cross-functional and intergovernmental crisis and consequence management response. ODP's national exercises and state and local domestic preparedness programs of exercises build on the office's training, technical assistance, and equipment program activities, and incorporate the tremendous instructional value of exercises into its Domestic Preparedness Program.

The National Exercise Program began in May 2000, when at the direction of the Congress, ODP conducted the TOPOFF (Top Officials) exercise, the largest federal, state, and local exercise of its kind, involving three separate locations and a multitude of federal, state, and local agencies. TOPOFF simulated simultaneous chemical, biological, and radiological attacks around the country and provided valuable lessons for the nation's federal, state, and local emergency response communities.

PROJECT RESPONDER

In determining the capabilities that emergency responders require, the committee examined in some detail the results of a national effort aimed at improving local, state, and federal emergency responders' capabilities to respond to the effects of terrorism-related weapons of mass destruction. Beginning in April 2001, well before the September 11 attacks, the National Memorial Institute for the Prevention of Terrorism in Oklahoma City, working initially with the Department of Justice and later with the DHS, contracted for a study to identify emergency responders' required capabilities and capability gaps. The study leveraged work conducted by the Interagency Board for Equipment Standardization and Interoperability (IAB) and worked with representatives of the first-responder community to identify requirements needed by the emergency responder community to mitigate the damage from a terrorist attack. The ultimate goals of this effort are to produce a national technology plan to help better focus research on the technological requirements of the responder community and to develop a Web-based, user friendly "responder knowledge base" of current and emerging systems for response to terrorism. This effort is called Project Responder.

The committee received a briefing[12] from the vice president of C4ISR for Hicks and Associates, Inc., one of the collaborators on Project Responder, and had access to the various reports developed by the project. This information enabled the committee to validate independently developed information against the Project Responder database. Tables 3-1 through 3-4 present information from

[12]Guy Beakley, Hicks and Associates, Inc., "C4ISR Requirements for the Nation's First Responders from Project Responder," briefing to the committee, Washington, D.C., July 22, 2003.

BOX 3-1
National Terrorism Response Objectives

Personal Protection
Detection, Identification, and Assessment
Unified Incident Command Decision Support and Interoperable Communications
Response and Recovery
Emergency Management Preparation and Planning
Crisis Evaluation and Management
All-Source Situational Understanding
Medical Response
Public Health Readiness for Biological Agent Events
Logistics Support
Criminal Investigation and Attribution
Agricultural Mitigation and Restoration

NOTE: Objectives in bold type relate to some aspect of C4ISR. SOURCE: Pollard et al. (2003).

Project Responder depicting some of the capability shortfalls as measured against several of the National Terrorism Response Objectives (see Box 3-1).

Detection, Identification, and Assessment

In the area of detection, identification, and assessment of WMD threats, the following specific C4ISR capabilities needed by emergency responders are identified:

- *On-scene detection:* Initial detection and characterization of danger to self and others; inclusion of detection before an event or onset of symptoms and characterization of suspicious objects;
- *Remote and standoff detection:* Identification and assessment of threat from outside the hot zone; remote sensors (e.g., lidar or directional/ imaging detectors), and/or point sensors mounted on robotic ground and air vehicles;
- *Detector arrays and networks:* Sensor arrays that can be networked to provide alerts, identification, and localization of chemical, biological, radiological, nuclear, and explosive threats; linkage to command data centers; provision of environmental monitoring in urban centers, building interiors;
- *Epidemiological surveillance and information systems:* Initial detection and characterization of a WMD event through public health and veterinary

surveillance; data-mining tools to detect abnormal levels of illness; linkage to suggested tactics, techniques, and procedures specific to the detected threat; and

- *Remote detection of deception/intent:* Noninvasive, noncontact detection of human deception and hostile intent at security checkpoints.

Table 3-1 presents the Project Responder capability assessment for the detection, identification, and assessment of WMD threats.

Unified Incident Command Decision Support and Interoperable Communications

In the area of unified incident command decision support and interoperable communications, the following specific C4ISR capabilities needed by emergency responders are identified:

- *Point location and identification:* The ability to know and visualize at all times the location and identity of individual responders, regardless of their position or movement;
- *Seamless connectivity and integration:* Communications systems that are able to seamlessly and dynamically interconnect multiple interagency users (with multiple functions) and information and communications technology systems;
- *Information assurance:* Guarantees of the availability, confidentiality, security, and integrity of information and information systems, including redundant systems;
- *Incident command information management and dissemination:* The ability to provide decision support, situation and resource status management, communications system management, and mission and task tracking; and
- *Multimedia-supported telepresence:* Provision of a multimedia telepresence between incident commanders, response personnel, technical specialists, and off-site facilities.

Table 3-2 presents the Project Responder capability assessment for unified incident command decision support and interoperable communications.

Emergency Management Preparation and Planning

In the area of emergency management preparation and planning for WMD scenarios, the following specific C4ISR capabilities needed by emergency responders are identified:

- *Risk awareness and assessment:* Assessment and analysis of threat, vulnerability, and criticality of events, venues, and systems (including key assets and infrastructures);
- *High-value target identification and monitoring:* Retention of the identity of high-value targets, use of appropriate monitoring techniques, communication of status whenever needed, and addressing of transitional threats; and
- *Disseminating threat and situation advisories:* Timely dissemination of vetted, evaluated, and actionable intelligence; audience-specific information; inclusion of local through national-level threat advisories.

Table 3-3 presents the Project Responder capability assessment for emergency management preparation and planning for WMD scenarios.

Crisis Evaluation and Management

In the area of crisis evaluation and management for WMD scenarios, the following specific C4ISR capabilities required for emergency responders are identified:

- *Threat assessment data collection and analysis:* The ability to collect specific and potential threat-related information, analyze the data, and validate and assess the threat for purposes of identifying the threat credibility;
- *Threat-relevant data dissemination:* The ability to identify what kinds of threat related information needs to be disseminated, identify who needs to receive what information, and deliver the right information to the right people; and
- *Tactical threat assessment:* The ability to assess threats inside buildings (i.e., "seeing" through walls), awareness of perpetrators' actions and of position and status of devices and weapons; risk, hazard, and situational size-up (quick assessment); and identification of individuals and objects that are at risk.

Table 3-4 depicts the Project Responder capability assessment for crisis evaluation and management for WMD scenarios.

Summary of Project Responder Capability Assessment

From the number of "red" entries (signifying "high risk; capability not currently available, fundamental science and technology work needed") in Tables 3-1 through 3-4, it is clear that many capabilities for emergency responders have not yet been met.

TABLE 3-1 Capability Shortfalls for Emergency Responders in the Detection, Identification, and Assessment of Weapons of Mass Destruction Threats

Capability	Chemical	Biological	Radiological	Nuclear	High-Explosive/ Incendiary
On-scene detection	Yellow	Red	Yellow	Red	Yellow
Remote and standoff detection	Red	Red	Red	Red	Yellow
Detector arrays and networks	Red	Red	Red	Red	Red
Epidemiological surveillance and information systems	Yellow	Yellow	Yellow	N/A	N/A
Remote detection of deception/intent	Red	Red	Red	Red	Red

NOTES: Red = High risk; capability not currently available, fundamental science and technology work needed. Yellow = Medium risk; technology exists but needs significant development. Green = Low risk; technology exists and simply needs maturation (none in this category in Table 3-1). N/A = Not applicable. SOURCE: Adapted from Guy Beakley, Hicks and Associates, Inc., "C4ISR Requirements for the Nation's First Responders from Project Responder," briefing to the committee, Washington, D.C., July 22, 2003.

TABLE 3-2 Capability Shortfalls for Emergency Responders in Unified Incident Command Decision Support and Interoperable Communications

Capability	Information Acquisition	Information Assessment and Course-of-Action Development	Decision Making	Direction
Point location and identification	Yellow	Red	Red	N/A
Seamless connectivity and integration	Red	Red	Red	Red
Information assurance	Red	Red	Red	Red
Incident command information management and dissemination	Yellow	Red	Red	Red
Multimedia-supported telepresence	Yellow	Yellow	Yellow	Yellow

NOTES: Red = High risk; capability not currently available, fundamental science and technology work needed; Yellow = Medium risk; technology exists but needs significant development; Green = Low risk; technology exists and simply needs maturation (none in this category in Table 3-2); N/A = Not applicable. SOURCE: Adapted from Guy Beakley, Hicks and Associates, Inc., "C4ISR Requirements for the Nation's First Responders from Project Responder," briefing to the committee, Washington, D.C., July 22, 2003.

TABLE 3-3 Capability Shortfalls for Emergency Responders in Emergency Management Preparation and Planning for Weapons of Mass Destruction Scenarios

Capability	Chemical	Biological	Radiological	Nuclear	High-Explosive/ Incendiary
Risk awareness and assessment	Yellow	Yellow	Yellow	Yellow	Yellow
High-value target identification and monitoring	Red	Red	Red	Red	Red
Disseminating threat and situation advisories	Yellow	Red	Yellow	Yellow	Yellow

NOTES: Red = High risk; capability not currently available, fundamental science and technology work needed; Yellow = Medium risk; technology exists but needs significant development; Green = Low risk; technology exists and simply needs maturation (none in this category in Table 3-3); N/A = Not applicable. SOURCE: Adapted from Guy Beakley, Hicks and Associates, Inc., "C4ISR Requirements for the Nation's First Responders from Project Responder," briefing to the committee, Washington, D.C., July 22, 2003.

TABLE 3-4 Capability Shortfalls for Emergency Responders in Crisis Evaluation and Management for Weapons of Mass Destruction Scenarios

Capability	Chemical	Biological	Radiological	Nuclear	High-Explosive/ Incendiary
Threat assessment data collection and analysis	Red	Red	Red	Red	Red
Threat-relevant data dissemination	Red	Red	Red	Red	Red
Tactical threat assessment	Yellow	Red	Yellow	Yellow	Yellow

NOTES: Red = High risk; capability not currently available, fundamental science and technology work needed; Yellow = Medium risk; technology exists but needs significant development; Green = Low risk; technology exists and simply needs maturation (none in this category in Table 3-4); N/A = Not applicable. SOURCE: Adapted from Guy Beakley, Hicks and Associates, Inc., "C4ISR Requirements for the Nation's First Responders from Project Responder," briefing to the committee, Washington, D.C., July 22, 2003.

REFERENCES

Canada, B. 2003. Homeland Security: Standards for State and Local Preparedness. May 12. Available online at <http://public.ansi.org/ansionline/Documents/Standards%20Activities/Homeland%20Security%20Standards%20Panel/RL31680.pdf>. Accessed November 20, 2003.

Cashin, T., D. Evans, and B. Salis. 2003. First Responder's Panel. Pp. 52–56 in Urban Security: Engineering the Protection of our Cities, Proceedings of the Conference on Urban Security: Engineering the Protection of Our Cities, October 7. George Bugliarello, ed. Brooklyn, N.Y.: Polytechnic University.

Collins, J.J. 2000. Training America's Emergency Responders: A Report on the Department of Justice's Center for Domestic Preparedness and the U.S. Public Health Service's Noble Training Center, Fort McClellan, Anniston, Alabama. July. Available online at <http://www.csis.org/homeland/reports/FirstResponders.html>. Accessed on September 25, 2003.

IOM (Institute of Medicine). 2002. Preparing for Terrorism—Tools for Evaluating the Metropolitan Medical Response System Program. Washington, D.C.: The National Academies Press.

ISTS (Institute for Security and Technology Studies). 2001. Cyber Attacks During the War on Terrorism: A Predictive Analysis. September 22. Available online at <http://www.ists.dartmouth.edu/ISTS/counterterrorism/cyber_a1.pdf>. Accessed September 25, 2003.

Jackson, B., D.J. Peterson, J. Bartis, T. LaTourrette, I. Brahmakulam, A. Houser, and J. Sollinger. 2002. Protecting Emergency Responders: Lessons Learned from Terrorist Attacks. Available online at <http://www.rand.org/publications/CF/CF176/>. Accessed September 24, 2003.

Larson, E.V., and J.E. Peters. 2001. Preparing the U.S. Army for Homeland Security: Concepts, Issues, and Options. Available online at <http://www.rand.org/publications/MR/MR1251/>. Accessed September 29, 2003.

LaTourrette, T., D.J. Peterson, J.T. Bartis, B.A. Jackson, and A. Houser. 2003. Protecting Emergency Responders, Volume 2: Community Views of Safety and Health Risks and Personal Protection Needs. Available online at <http://www.rand.org/publications/MR/MR1646/>. Accessed August 21, 2003.

NIJ (National Institute of Justice). 2003. When They Can't Talk Lives Are Lost: What Public Officials Need to Know About Interoperability. February. Available online at <http://www.agileprogram.org/ntfi/ntfi_brochure.pdf>. Accessed April 1, 2004.

NRC (National Research Council). 2003. Science and Technology for Army Homeland Security, Report 1. Washington, D.C.: The National Academies Press.

OHS (Office of Homeland Security). 2002. National Strategy for Homeland Security, July. Available online at <http://www.whitehouse.gov/homeland/book/nat_strat_hls.pdf>. Accessed September 24, 2003.

Pollard, N.A., R.V. Tuohy, and T. Garwin. 2003. Project Responder Interim Report: Emergency Responders' Needs, Goals, and Priorities, March. Oklahoma City, Okla.: The Oklahoma City National Memorial Institute for the Prevention of Terrorism.

Schwabe, W., L.M. Davis, and B.A. Jackson. 2001. Challenges and Choices for Crime-Fighting Technology: Federal Support of State and Local Law Enforcement. Available online at <http://www.rand.org/publications/MR/MR1349/>. Accessed September 26, 2003.

U.S. Congress. 2002. First Responder Terrorism Preparedness Act of 2002, Senate Report 107-295, October 1. Available online at <http://frwebgate.access.gpo.gov/cgi-bin/useftp.cgi?IPaddress=162.140.64.21&filename=sr295.pdf&directory=/diskb/wais/data/107_cong_reports>. Accessed September 24, 2003.

4

Defense Technologies for Homeland Security

INTRODUCTION

Overview and Scope

This chapter focuses on the technologies that are currently being developed in the Army or other components of the Department of Defense (DOD) in the area of command, control, communications, computers, intelligence, surveillance, and reconnaissance (C4ISR) and which the committee believes may have *potential* application to the homeland security mission and emergency responders. Much of the information contained in the chapter is derived from Army and DOD documentation, from briefings presented to the panel, and from first-hand knowledge of the study committee members. No attempt is made to offer a comprehensive presentation with respect to these technologies, because of both space and study schedule limitations. Rather, it is the committee's intent to present to the Army and the homeland security community those technologies that the committee believes may have relevance for emergency responders and which could prompt further interaction between the Army and the emergency responder community.

Very little discussion of commercial programs is presented here, as the committee believed that to do a credible and comprehensive job in such an endeavor would far exceed the scope of the present report; also, it was reluctant to highlight a particular commercial product without reviewing other similar available products. Nevertheless, there certainly are products being developed in the commercial world that would be of great benefit to the emergency responder. A major contribution has been made by commercial industry in the development of software tools, particularly decision-making tools, that can easily be adapted to military and/or emergency responder use. Likewise, commercial standards such as those established by the Institute of Electrical and Electronics Engineers and current and evolving Internet protocols can be very helpful in achieving interoperability across the plethora of agencies involved in homeland security, and the committee believes that these standards ought to be categorized and incorporated into any equipment development programs by the Department of Homeland Security (DHS).

Conclusion 4-1. The U.S. Army has developed a significant number of C4ISR technologies for the Future Force that appear to have direct applicability to the emergency responder community.

Recommendation 4-1. The U.S. Army and the Department of Homeland Security should evaluate the systems described in Chapter 4 of this report for their potential to support interagency collaboration.

Organization of This Chapter

Following the methodology adopted in Chapters 2 and 3, the committee divided the C4ISR elements as follows: command, control, and computers (C3); communications (C); and intelligence, surveillance, and reconnaissance (ISR). As explained previously, the choice of command, control, and computers as a grouping was made because of the integral nature of decision-making algorithms and software now so prevalent in command-and-control systems, and in fact enabled by the vast data capacity and fast processing made available by today's computers. "Communications" stands by itself as the backbone of any such system. The ISR aspect of C4ISR is treated as a single entity because of the overlapping technologies underpinning the area.

This chapter begins with a general discussion of the technical issues associated with C4ISR, primarily from a broad system perspective. The committee identifies some broad-based programs and tools, at the integrated system level, that may be of interest to emergency responders and to the DHS. After a general technical discussion of C4ISR, attention is turned to the component-level areas (C3, C, and ISR). Finally, after a discussion of the technologies, the committee identifies some of the programs believed to be relevant, perhaps with modifications, to the emergency responder community and the DHS. Tables throughout this chapter (Tables 4-1 through 4-5) summarize the technologies relevant to each major aspect of C4ISR.

Finally, the committee offers some comments on other programs and activities within the DOD that, although outside the strict C4ISR arena, offer real value to the emergency responder. For example, attention is given to the major investment and significant advantages available in the DOD modeling and simulation arena.

C4ISR Technical Description

General Description

Science and technology (S&T) in the area of C4ISR is designed to enable comprehensive situational awareness for network-centric operations (U.S. Army, 2003). As such, it has several technical components: the generation of sensor data; the processing required to turn these sensor data into information; the movement of the data or information through a communications system to another location; the integration of the information from various sources, both internal and external, to turn it into intelligence; the presentation of the intelligence to a user or a decision-support software program in a comprehensible fashion; and the dissemination of the decisions (commands) and selected information to subordinate elements. Each activity described here has its own unique—and sometimes complex—technology.

Integrated System Issues

There are several issues associated with an effective C4ISR system that must be addressed in any technical solution. Many of these issues are being addressed in current Army and DOD programs. One such issue is whether or not the system will rely on commercial infrastructures, particularly power and fixed communications lines, or will be entirely self-reliant. Another key issue is mobility—that is, whether the system is fixed at one or more locations or is mobile. Of importance are the power and bandwidth issues, particularly those involving high-bandwidth video imagery, associated with the sensors and the movement of raw or processed data. A key parameter is information latency, tied closely to computer processing time and communication bandwidth. Another sometimes-related parameter involves the uncertainty associated with the generation of information and how the information will be used in the decision-making process. Another major systems issue is cost, particularly life-cycle cost. A brief discussion of some aspects of these key parameters follows.

Reliance on existing infrastructure versus a completely independent system as the backbone or even a component of a C4ISR system is probably the key design issue for the entire system, particularly with respect to the sensor system and the communications system. The design of a sensor system may be entirely different if it is a mobile, deployable system relying on battery power and a limited communications system—such as the Army's legacy Single Channel Ground and Airborne Radio System—instead of being at a fixed site, utilizing infrastructure power and communication lines. In the latter case, power and bandwidth issues are not as constraining as in the former case, and even in the event of a power failure, there is usually a redundancy in power that will keep the system up and running for a period of time. However, in a catastrophic event, emergency responders may not be able to rely on fixed infrastructure, and so the capability of an independent system, at least on a temporary basis, would be prudent.

Mobility issues are related to but not identical to fixed-infrastructure issues. A mobile system can still rely heavily on the use of commercial power or communication lines as part of its technical solution. A completely independent mobile system using its own power and communications infrastructure is probably the ideal case—although, owing to power and bandwidth constraints, there may be limitations on system performance. The Army's Warfighter Information Network-Tactical (WIN-T) and C3[1]-on-the-Move Demonstration are addressing both the independent infrastructure and the mobility issues.

Data latency (i.e., how much and how fast data or information can be transmitted across a network) is very important. In some cases, particularly if an emergency responder's life may be at risk, late information or direction may be useless or even detrimental if the responder was relying on it. Latency is associated with processing power (how fast sensor data can be converted to information) and communication bandwidth availability (how fast the information can be provided to the appropriate decision maker or user). Closely related to this issue are the generation and transport of video imagery across the network. Video imagery requires considerable bandwidth, raising the question of what is good enough. Are two or three frames sufficient, or does the military or incident commander need full video? If the latter, what frame rates are sufficient—say, 10 frames per second or 30 frames per second? Is a sensor processor declaration that an object is a T-72 tank sufficient, or must the analyst see an image of it to be sure? The Army's Network Sensors for the Future Force Program is addressing the generation and transport of video imagery across the network.

The uncertainty in information and the conversion of information to intelligence are also key issues, not independent of the latency and amount of information transferred. Uncertainty is generated in different ways. It could arise when a sensor, using automatic target recognition software, identifies a T-72 tank with a 90 percent confidence interval. It could also result from a human looking at an image and making a best guess at identifying it. Using aggregated information and trying to infer enemy composition and intent involve an inherent uncertainty that must be understood when courses of action are developed. The Army's Knowledge Fusion Program is looking at some of these issues.

Finally, there is the difficulty of delivering information to the end user as it is needed and in a form that supports the task at hand, whether the user is a civilian incident commander, an infantry squad leader, or an individual soldier. Excess information, causing information overload, can be as detrimental as too little information. The concept of "push-pull" has been used in this area, with "push" meaning that certain information deemed to be important to a certain user is

---

[1]In this case, C3 retains the conventional meaning of command, control, and communications.

automatically sent to him or her based on some pre-selected criteria, and "pull" meaning that information is sent only when it is asked for. How the information is presented, whether on a computer display, in an audio message, by a vibration, and so on, is also important.

Additionally, the problem of cost is associated with some of the Army's high-performance sensors, particularly high-resolution, platform-based infrared imaging sensors. No matter how good they are, if the sensors cost too much, the system can become unaffordable for the Army. Cost may also be important to emergency responders because much equipment is procured through local budgets or grants. If emergency responder acquisitions cannot be bundled to achieve the economies of scale seen in military procurements, costs may be prohibitive at the local level.

Relevant Integrated Systems Technology Programs

Several programs attempt to look at the C4ISR system as a whole and to deal with the full complexity of systems integration. Three of these programs are noted here, with one, the C3-on-the-Move Demonstration, looking at providing integrated information to the rapidly moving platforms associated with Future Force and Future Combat Systems (FCS), and the other two, Land Warrior and Future Force Warrior, focusing more at the lower-echelon, infantry-level platform, the soldier. The committee believes that these programs are of particular importance for their applicability to emergency responders. The relevant integrated systems technology programs are shown in Table 4-1.

C4ISR COMPONENT TECHNOLOGIES AND PROGRAMS

Command, Control, and Computer Technologies

As discussed above, C3 technologies can support emergency responders' need for informed event management. The following is a general discussion of technical issues and a follow-on discussion of applicable DOD programs. Table 4-2 presents information on the relevant C3 technologies, including a brief description or statement of purpose and an availability assessment.

General Discussion of Technical Issues

The general technical issues associated with command and control are primarily focused on the aggregation of different information from various sources to support the incident commander's decision-making process. The ability to fuse information from different sources into a coherent picture of the battlespace or disaster scene and the use of decision-support tools are the key technical aspects.

TABLE 4-1 Integrated Systems Technology Programs Relevant to Emergency Responders

Program	Description	Availability[a]
Command, Control and Communications on-the-Move Demonstration (Army S&T)	Demonstration of an integrated C3 on-the-move capability utilizing intelligence, surveillance, and reconnaissance assets and networked firepower,[b] which will show that the information from these sensors can be moved to a command-and-control location, on the move, digested, and disseminated by a command-and-control system quickly and effectively (Fillian;[c] U.S. Army, 2003). Technologies from this effort can assist responders in developing more efficient mobile command centers.	R
Land Warrior (LW) Program (Army Acquisition)	Program designed to significantly improve the capability of the individual soldier and to implement the soldier-as-a-system concept. One of its key elements will be improved C4ISR. It integrates many commercial and government off-the-shelf technologies into the soldier platform. It combines computers, lasers, geolocation, and radios with the soldier's current mission package, giving him or her a significant increase in C4ISR capability. The program is structured in three phases: (1) LW-IC, or initial capability in FY 2004; (2) LW-SI, or Stryker Interoperable, providing recharge on the move and expanded situational awareness; and (3) LW-AC, or advanced capability, incorporating several improvements from the Army's Future Force Warrior program: specifically, weight reduction and extended mission duration (U.S. Army, 2003). Technologies from this program will enhance the capabilities of individual responders. For example, a firefighter in a large, smoke-filled building would have better awareness of his or her own location and that of fellow responders, as well as access to critical information. The firefighter would also have better communications with his or her team and leaders.	N
Future Combat Systems (FCS) C4ISR	Secure C4ISR system to harness advances in the distribution and effective use of information power (U.S. Army, 2003). The FCS C4ISR programs are a component of the FCS program currently managed via a Lead System Integrator contractor.	N
Future Force Warrior (FFW) Advanced Technology Demonstration (Army S&T)	Effort to allow the individual to interface with external platforms and sources of information, including unmanned aerial vehicles, unmanned ground vehicles, and the Future Force C4ISR network. It will integrate the Joint Tactical Radio System squad-level communications system (described in Table 4-3), allowing an interface with the	F

TABLE 4-1 Continued

Program	Description	Availability[a]
	integrated force structure, and provide information to support networked firepower. Another key improvement, particularly of interest to emergency responders, is the Warfighter Physiological Status Monitor, allowing the commander to track the health status of the individual soldier at all times. This system allows the commander to dispatch medical assistance whenever necessary. The FFW program is also addressing the two critical issues associated with these individual, mobile systems—power and weight. These two parameters are the real limitation to any such system, and new power sources, such as fuel cells, and lightweight materials are the potential answers (U.S. Army, 2003). Individual responders will have even more enhanced capabilities than those from Land Warrior technologies. Not only will responders be able to operate more efficiently individually, but they will also be more effective as a team. Leaders will also have better awareness of the status of the responders under their control.	

[a]Availability: R, ready (TRL 8-9); N, near term (TRL 4-7); F, far term (TRL 1-3). See Appendix G in this report for descriptions of technology readiness levels (TRLs).

[b]"Networked firepower" means the coordinated use of a variety of munitions such as artillery, rockets, and so on.

[c]Larry Fillian, Director, Command and Control Directorate, Communications-Electronics Research, Development and Engineering Center, "C4ISR Enabling Technologies," briefing to the committee, Washington, D.C., July 22, 2003.

TABLE 4-2 Summary of Programs Relevant to Emergency Responders: Command, Control, and Computer (C3) Technologies

Program	Description	Availability[a]
Command and Control in Complex and Urban Terrain (Army S&T)	A suite of command-and-control tools for the dismounted warrior in an urban environment, providing enhanced collaboration, information reach-back, mixed asset management, and seamless situational understanding. In particular, this program will develop distributed command-and-control tactical decision aids, applications, and models addressing decision making with partial and missing information in complex/urban terrain (U.S. Army, 2003). Tools from this effort will allow responders to better manage personnel and equipment assets in an urban environment. In particular, responders will have a better view of the urban situation and will be able to make more informed decisions.	F

continued

TABLE 4-2 Continued

Program	Description	Availability[a]
Battle Terrain Reasoning and Awareness (Army S&T)	A comprehensive suite of terrestrial and lower-atmosphere battlespace environment tactical decision aids (TDAs) that generate information and knowledge necessary to enable decision and execution processes across C4ISR systems. These tools capture the interrelationships and effects of terrain and weather on force/threat behavior as well as platform and system performance. TDA-generated information and knowledge products will be of robust content and lightweight structure, supporting tactical dissemination and automated decision support tools of other C4ISR system-specific C4ISR decision-support tools (U.S. Army, 2003). Tools from this effort will greatly support responders after a hurricane or other natural disaster, during which large regions have lost power and communications. Responders will be able to analyze terrain and weather data to best determine the location of communication and sensor assets.	F
Geospatial Information Integration and Generation Tools (Army S&T)	Tools to integrate, manage, and exploit multisource data imagery, features, and elevation data to present only relevant terrain data to the user. Work will be done to fuse the data from synthetic aperture radar, inverse synthetic aperture radar, infrared, and other sender data into digital terrain maps. Algorithms will be developed to assist in feature extraction, automatic determination of optimal routes of movement in and out of an area, automatic damage assessment, and integration of satellite data. The end use for responders should be the ability to reach out over a network and retrieve the latest, multisenor, multiphenomenology terrain and feature data available and to use intelligent tools to assist in the processing and evaluation of the information presented (U.S. Army, 2003).	F
Agile Commander Advanced Concept Technology Demonstration (ACTD) (Army S&T)	A dispersed, highly mobile command post that provides the commander with continuous, responsive, proactive, real-time battlespace management information during both stationary and mobile operations. The Agile Commander will provide a scalable and reconfigurable command, control, computers, communications, and intelligence multifunction operator environment with access to all command post information. One of the key tools being developed is the Distributed Analysis and Visualization Infrastructure (DaVinci) tool set. This tool set is an advanced suite of decision aid software that executes execution-centric, mobile command and control (U.S. Army, 2003). This effort can be leveraged by responders to develop better mobile command centers.	F

TABLE 4-2 Continued

Program	Description	Availability[a]
Homeland Security Command-and-Control Advanced Concept Technology Demonstration (Defense Information Systems Agency and U.S. Air Force [USAF] S&T)	A 5-year program to define, refine, and transition technologies and concepts of operation to significantly increase DOD homeland security responsiveness in areas of consequence management, crisis response, deterrence, and intelligence coordination. The assured communications must be deployable, flexible, redundant, wireless, and protected. The interoperability capability must use hardware and software that operate across all levels of government under daily conditions, conditions of increased vigilance, and crisis. The threat alerts/attribution capability must focus on prediction, alerts, warnings, and prevention, as well as pattern and relationship identification. The command, control, and communications portion will focus on the full range of capabilities to plan, assess, make decisions, communicate decisions, and receive feedback.[b] This effort will significantly improve the interoperability between the military and responders.	N
Knowledge Fusion (Army S&T)	An effort to resolve the main problem with nascent knowledge management systems that overload the user with information. Intelligent agents are used to break large problems into smaller components that can (possibly) be handled in a parallel manner. Ontology agents identify classes of information and organize them hierarchically according to user-established rules.[c] The integration of similar intelligent agents in responder decision-making tools will reduce the detrimental effects of information overload.	F
Joint Blue Force Situational Awareness (JBFSA) ACTD (Office of the Secretary of Defense S&T)	Software interfaces and connectivity enabling the integration of existing blue force tracking systems to create a blue force situational awareness picture within the global command-and-control system family of systems common operational picture. JBFSA ACTD will provide improved situational awareness, tracking, tagging, and locating, as well as logistics and asset management information to the Joint Force commander's common operational picture (DOD, 2003). The employment of similar software interfaces in responder command-and-control systems will help provide leaders a more integrated view of their personnel and equipment assets with respect to the emergency situation.	N

continued

TABLE 4-2 Continued

Program	Description	Availability[a]
Future Command Post Technologies (Army S&T)	Prototype command-and-control product applications for functionally and physically agile, rapidly deployable, and distributed operations that will enable commanders to execute operations ranging from war to humanitarian assistance. The technical integration and development effort includes command-and-control tools and mobile adaptive computing.[d] This effort can be leveraged by responders to develop better command centers.	F
Intelligent Information Technology (Defense Advanced Research Projects Agency and USAF S&T)	An effort to enable the military/emergency crisis responder team to rapidly obtain and assimilate information and knowledge relevant to the decisions that must be made in an ongoing crisis or conflict situation. It will develop and demonstrate new technology to detect and identify the presence of biowarfare or bioterrorist attack; rapidly build and use comprehensive knowledge bases to interpret, reason, and respond to the changing critical situation; and develop multimodal human identification (HumanID) biometric technologies to detect, recognize, and identify humans at a distance to support early warning, force protection, and operations against terrorist, criminal, and other human-based threats.[d] Tools from this effort will greatly assist responders in accessing military and civilian information and knowledge in the areas described in this entry.	F
Forecasting, Planning, and Resource Allocation (U.S. Navy [USN], USAF, Army S&T)	Secure, network-centric, intelligent-agent-assisted collaboration environment for faster decision making. It will demonstrate intelligent, self-organizing, adaptive, agent-based software allowing commanders to interactively create, share, and merge plans; monitor execution; and interactively repair plans.[d] The integration of similar intelligent-agent-assisted software into responder command-and-control systems will greatly enhance the capability of responder leaders to manage a situation.	F
Decision Support Systems for Command and Control (USN S&T)	Technologies to enhance the decision-making skills of military commanders and their battle staffs. Example technologies include computational models of human information processing and decision making, Bayesian models for effects-based planning, and advanced multimodal workstations for decision-support systems.[d] The integration of similar models and multimodal workstations into responder command-and-control systems will greatly enhance the capability of responder leaders to manage a situation.	F

TABLE 4-2 Continued

Program	Description	Availability[a]
Commander-in-Chief (CINC) 21 ACTD (USN and Office of the Secretary of Defense S&T)	Technologies to enhance the efficiency and coordination of joint and coalition operations through decision-focused command-and-control support functions; enhance the speed and quality of command decision making through the exploitation of knowledge and information management tools and the identification, extraction, and optimal presentation of knowledge and information to decision makers; improve the ability of the CINC's "extended" staff to track and manage multiple simultaneous crises; and free decision makers from being tied to their command centers (AITS-JPO, 2003). Tools from this effort should be leveraged to enhance responder command-and-control systems, especially in the areas of decision-making processes and the tracking and managing of crises.	N

[a]Availability: R, ready (TRL 8-9); N, near term (TRL 4-7); F, far term (TRL 1-3). See Appendix G in this report for descriptions of technology readiness levels (TRLs).

[b]Glenn Cooper, Assistant Technical Manager, Defense Information Systems Agency, "Homeland Security/Homeland Security Command and Control ACTD," briefing to the committee, Washington, D.C., August 25, 2003.

[c]Dan Kuderna, Communications and Electronics Research, Development and Engineering Center, "Fusion-Based Knowledge for the Objective Force," briefing to the committee, Washington, D.C., August 26, 2003.

[d]Selected information provided by the Office of the Director, Defense Research and Engineering, December 4, 2003.

Fusion Technologies. Information fusion is the combination and distillation of information from various databases driven by a set of search algorithms designed to focus on answers to a set of queries. Knowledge management, expert systems, and artificial intelligence all contribute to information fusion. Knowledge management combines the capture of an organization's information with (relatively) easy retrieval and use of that information by the corporate body. The intent is to make the knowledge that exists in a variety of locations available to the entire organization. Expert systems and artificial intelligence generally focus on narrow domains to assist human endeavors. Fusion technology systems attempt to take functions performed by humans and assist or replicate the actions of the human.

Image fusion combines images from several or various types of sensors into a single image for the viewer. This image, for example, could take the form of a digital terrain map that has icons of friendly and enemy forces depicted as an overlay. The data on the force locations could be received in message format from an intelligence or headquarters organization. It could also take the form, for example, of a combination of a photographic image taken in the visible spectrum, an infrared photo, and a synthetic aperture radar image. In both cases the images need to be "registered" to a common map grid so that, in the latter case, similar terrain features (such as a crossroad) appear in the same place on the resulting fused image.

Decision-Support Tools. As sensors proliferate and more and more information is provided to the incident commander, he or she can quickly become overwhelmed. Decision-support tools—many of which can be adapted from the commercial market—can help the incident commander to use this information wisely. These tools can be adapted to military or police tasks to indicate potential courses of action, such as showing how resources can be allocated and where deficiencies might arise.

Communications

General Discussion of Technical Issues

The history of the advance of radio communications shows an oscillation between commercial and traditional military applications. Table 4-3 presents information on communications technologies relevant to emergency responders, including a brief description or statement of purpose and an assessment of availability. The most recent major advances have come in the commercial sector, with the goal of giving mobility to telephone and Internet users. However, the goals and constraints of commercial and military radio systems are different. While both applications seek mobility, only the military seeks independence from fixed terrestrial infrastructure.

The goals and constraints of communications systems intended for homeland security and emergency workers lie somewhere between those of commercial systems and traditional military applications. Mobility is of course essential. However, depending on the size and nature of anticipated events or attacks, there can be some degree of reliance on fixed infrastructure. In order to discuss these differences in goals and constraints in greater depth, it is convenient to consider command and tactical networks separately.

Cost is an important consideration for the communications networks of emergency workers. Local governments (funding personnel such as firefighters and police) and companies with emergency workers (such as utilities) may not be willing to spend additional amounts for functionality that is used only for rare but large-scale emergencies. Thus, it may be important to engineer communications systems that can be upgraded incrementally. For example, a state or federal agency could bring radio equipment that would connect individual groups of emergency workers responding to large events.

Bandwidth is an increasingly scarce commodity, but there are work-arounds. Today, for example, 1000 simultaneous two-way conversations can be easily

TABLE 4-3 Summary of Programs Relevant to Emergency Responders: Communications

Program	Description	Availability[a]
Joint Tactical Radio System (JTRS) (Army Acquisition)	Software-reprogrammable, multiband/multimode-capable, network system that provides simultaneous voice, data, and video communications in order to increase interoperability, flexibility, and adaptability in support of varied mission requirements. This radio system will provide common, multimedia communications capabilities across all facets of the emergency responder community. It will also support the interface of civilian responders with military organizations.	R
JTRS Squad-Level Communications (Army S&T)	Effort focusing on the individual soldier and associated programs such as the Land Warrior Program and its follow-on program, the Future Force Warrior. Much of the technical basis of this program will be taken from the Small Unit Operations Situational Awareness System of the Defense Advanced Research Projects Agency. This radio system will enhance the multimedia communications capabilities of individuals and small teams of emergency responders.	F
Adaptive Joint C4ISR Node (Army S&T)	Demonstration of communication relay and signals intelligence/electronic warfare (SIGINT/EW) capability in a multifunctional, modular, scalable, and reconfigurable airborne payload. The primary function of the payload is to relay multiple types of communications waveforms, but it also provides SIGINT capability and information warfare capability. The system will provide connectivity and interoperability between disparate radios and networks (legacy and future, joint and coalition). This airborne system will enhance the interoperability of communications systems across the entire emergency responder community and with military organizations.	F
Warfighter Information Network-Tactical (WIN-T) (Army Acquisition)	Integrated, high-speed, and high-capacity backbone communications network for the Future Force, optimized for offensive and night operations and supporting multiple simultaneous missions. It is a tactical mobile network, based on commercial standards, leveraging the JTRS. It encompasses network infrastructure (integrated or embedded switching, routing, and transmission systems), network operations (naming, addressing, and user profiles), and user interfaces that support video, voice, and data transmission across the battlespace. This network will	N

TABLE 4-3 Continued

Program	Description	Availability[a]
	provide the communications backbone that emergency responders will need to adequately implement the enhanced command, control, computers, communications, intelligence, surveillance, reconnaissance systems identified in this report.	
Multifunctional on-the-move Secure Adaptive Integrated Communications Advanced Technology Demonstration	Effort to enable on-the-move network communications for the mobile, dispersed force in the Future Force concept. The focus of the program will be to integrate a highly adaptive communications infrastructure to support the seamless flow of multimedia services across terrestrial and space-based platforms. Its wireless communications architecture will support multimedia applications, quality of service for mobile networks, adaptive and ad hoc mobility protocols, bandwidth management, and both horizontal and vertical handoff in a mobile, wireless environment. This communications infrastructure will greatly enhance the mobile C4ISR capabilities of responders. This will be of great importance in situations in which local communications and power have been destroyed or shut down.	N

SOURCE: U.S. Army (2003).

[a]Availability: R, ready (TRL 8-9); N, near term (TRL 4-7); F, far term (TRL 1-3). See Appendix G in this report for descriptions of technology readiness levels (TRLs).

accommodated in a 25-megahertz (MHz) band. With new technologies, such as Wideband Code Division Multiple Access, commercial systems might handle closer to 2,500 simultaneous two-way conversations. Even greater capacity may be on the horizon (Ericsson, 2001). Accounting for dead time in each conversation will allow nearly doubling the number of conversations that can be accommodated. At an event covering a large area, frequencies can be reused in different parts of the site, allowing for even more conversations. Compressed video of standard quality can be sent at 30 frames per second with a data transmission speed of 6 megabits per second (Mbps), which can be accommodated in less than 6 MHz of bandwidth. A lower frame rate or reduced image quality may be acceptable, thereby reducing the bandwidth required for transmission. Cable modems give fast access to the Internet at about 1 Mbps, requiring 1 MHz bandwidth or less. Since the bandwidth for data access is needed only for short periods of time, many terminals can have access using the same frequency band.

Thus, bandwidths on the order of 25 MHz should be adequate to support the communications needs at a major crisis event. While public safety bands of 25 MHz exist, they are currently divided among various agencies in small disjointed segments. Each agency may deploy its segment of the bandwidth in a way that makes it impossible to share bandwidth with others, so no one can get the full advantage of its use. Developing a consensus among emergency workers to reform the use of bandwidth will require considerable resources and leadership.

Networks for Command Functions

A command network for vertical and horizontal communications among the commanders of different groups of emergency workers, with reach-back to local, state, and national headquarters, must accommodate voice, video, and data. In order to recover from information loss and to permit a forensic study of a crisis event, such a network should allow for recording and archiving the tactical communications at a fixed headquarters. For routine events, radio communications could be transmitted via terrestrial infrastructure designed to give good coverage over the region being served. However, to be prepared for a major catastrophic event involving destruction of the terrestrial infrastructure, some mechanism is needed for connecting to the fixed network via satellite, airborne relay, or terrestrial relay.

The functionality of this radio command network would be comparable to that envisioned for Joint Tactical Radio System (JTRS) Cluster 1, although it will not have the capability to work with legacy military systems (U.S. Army, 2003). In particular, the system could make use of the wider bandwidth available in some public safety bands and avoid the costly features of JTRS Cluster 1, which is software programmable and multiband. The command network must have access to databases in the fixed network and therefore can make use of Internet protocols, as is intended for Army networks. While the databases must be secure from malicious intrusion, it may not be necessary to encrypt transmissions over the networks.

Tactical Networks

Tactical networks of man-pack radios are needed to provide communications among incident commanders and individual emergency responders and among responders within a group. This network will be similar in use to that envisioned for JTRS Cluster 2, Squad-Level Communications, which at this time is in the research stage. Because this radio system is only in the research stage and is least influenced by legacy systems, it is well positioned to accommodate the needs of emergency responders. Moreover, fielding such a family of radios to all emergency responders could provide for interoperability among different groups of emergency responders, at least at the tactical level.

Emergency responders need radios capable of two-way voice and data transmission, possibly including low-rate video from helmet-mounted cameras. It is easy to envision many uses for the down-link transmission of data from the global information grid (GIG) (e.g., vehicular records, photographs) to the personal digital assistances (PDA)s of individual emergency responders, and digital up-link transmissions (requests for data, vital signs, and equipment status). This capability is consistent with the applications seen by the Army for squad-level communications.

Ad hoc network technologies are being considered for JTRS Cluster 2 in order to create robust, network-centric communications. Firefighters operate in small units and have communications procedures and needs similar to those of an Army squad. Communications links must provide coverage over large buildings, in tunnels, and in other difficult environments not conducive to signal propagation. Current approaches to providing such coverage are based on the use of repeaters carried by firefighters or fixed in the infrastructure. In the case of ad hoc networks, nodes brought in for an operation or permanently fixed in the infrastructure would replace the repeaters. With sufficient deployment of such nodes, good coverage could be obtained, even in difficult radio environments.

Under normal conditions, police and emergency medical services communicate with a dispatcher rather than with each other. In this case robust coverage may be achieved using an ad hoc network design with the additional fixed nodes distributed in the infrastructure, as discussed above.

Concerns

Given the scope of this study, the committee did not conduct a technical review of any of the communications programs listed in Table 4-3. However, on the basis of briefings provided to it and the knowledge of its members, the committee expressed some concerns with regard to the basic structure of the JTRS. The committee sees the program as being confronted with the challenge of retaining compatibility with legacy systems, while at the same time making significant progress toward the network-centric operations envisioned for the future battlefield. If this dilemma is not managed carefully, the outcome could be that of perpetuating the legacy radios at the expense of developing the low-cost, adaptable radio needed for the future. Additionally, while this type of radio would seemingly be very useful for the emergency responders in a local jurisdiction, such as the police force, firefighters, and National Guard and FBI personnel, the DOD requirements are well beyond what the local emergency responder needs and can afford.

The committee believes that the fundamental objective of future radio programs should be to implement the radio component of the network-centric warfare envisioned in the current DOD doctrine. To be successful, advantage must be taken of advances in commercial communications. The adoption of commercial standards should be an objective. In fact, a previous study by the National Research Council urged the military's participation in setting commercial standards to increase the commercial off-the-shelf (COTS) components of military radio systems (NRC, 1997). The military network solution must address problems such as routing in a mobile environment and the determination of adaptive and efficient ways to use the available spectrum. Addressing these problems will require taking advantage of the significant technology advances made by the cellular industry—which would best be done through the substantial involvement of that industry. This approach will also be necessary to enhance the radio in a straightforward manner and to produce it at a cost that would make it accessible to the civilian emergency responder community.

The future radio will require new modulation techniques, compression mechanisms, and error-correction features. These elements are clearly in the radio domain. Other functionality such as communications security, routing, and gateway services could be allocated to the radio or the terminal. In the end only certain pieces will be designated as radio components. Therefore, the tactical radio of the future must be designed in the context of the overall architecture envisioned for conducting network-centric information warfare.

The future radio is a software-defined radio (SDR). SDR technology can be used in any device that uses radio frequencies for communication, including cellular base stations, military communications systems, and public safety radios. SDR appears to be the best approach for interoperability, since it provides an immediate, cost-effective solution that does not require organizations to purchase new radios. A portable SDR device brought to an emergency scene can enable interoperability among selected members of different agencies by creating communication links between different radios and establishing infrastructure where none exists, or supplementing inadequate existing infrastructure by serving as a base station (Steinheider, 2003).

The DOD recognizes the potential efficiency and performance of SDR and has established the Joint Tactical Radio System Joint Program Office (JTRS JPO) to pursue this technology. The JPO has begun to acquire software implementations of a first set of 33 communications standards. The key standard is the software communications architecture (SCA), intended to ensure interoperability across platforms from many vendors. Thus, the JTRS operational requirements document (ORD) identifies compatibility with many military and commercial waveforms (JROC, 2003). The SCA standardizes the software's operating environment, as well as the control and communications mechanisms for both the hardware and the external interfaces of the radio. Many NATO allies have signed agreements to apply the SCA in future acquisitions, and the JPO hopes that the SCA will become the basis for commercial SDR software standards (Steinheider, 2003).

Intelligence, Surveillance, and Reconnaissance

General Discussion of Technical Issues

Generally speaking, the technologies associated with ISR are the processing technologies at sensors or the sensor nodes, sensor communication networks that carry either the raw data or the processed data or information, higher-level fusion (discussed above) and processing aids such as automatic target recognition or higher-level aggregation and interpretation software, and displays. Table 4-4 presents information on the relevant ISR technologies, including a brief description or statement of purpose and an availability assessment.

TABLE 4-4 Summary of Programs Relevant to Emergency Responders: Intelligence, Surveillance, and Reconnaissance (ISR)

Program	Description	Availability[a]
Joint Intelligence, Surveillance and Reconnaissance Advanced Concept Technology Demonstration (ACTD) (Office of the Secretary of Defense [OSD] S&T)	Provide timely top-down/bottom-up information to enable enhanced battlespace visualization. The objective is to provide a significantly enhanced capability to dominate situational awareness through the use of a Web-based browser and information agents, Joint Technical Architecture-compliant sensor interfaces, commercial/government off-the-shelf complexity reduction tools, distributed database management, and improved visualization and display tools (U.S. Army, 2003). This effort will greatly enhance the ISR capabilities of emergency responders at the regional, state, and national levels. It will also assist these same responders in accessing ISR information from military and national assets.	N
Networked Sensors for the Future Force[b] (Army S&T)	Develop and integrate off-board sensor packages onto mobile platforms (unmanned ground vehicles [UGVs], mini unmanned aerial vehicles [UAVs], unattended ground sensors [UGSs]) and create a system of systems that can be networked in complex terrain (including urban areas). The program integrates and demonstrates enabling sensors—uncooled infrared (IR), flash laser with short-wave IR, mini UAV, UGV, microsensors (acoustic, seismic, IR imaging, magnetic, and radio frequency); assesses network communications performance resulting from the Warrior Extended Battlespace Sensors science and technology objective (STO) and Smart Sensor Communications Network STO; and includes an intelligence reach-back capability for threat profile development, sensor deployment, and smart data management (U.S. Army,	F

TABLE 4-4 Continued

Program	Description	Availability[a]
	2003). This networked sensor system will enhance ISR capabilities of emergency responders at all levels, but especially at the individual and small-team level.	
Advanced Night Vision Goggle (ANVG) (Army S&T)	Develop and demonstrate the Air Warrior operations requirement for an integrated, 100-degree field of view helmet-mounted night vision goggle system. The ANVG will be a modular horizontal technology integration design that can also meet requirements for Mounted Warrior and Land Warrior, allowing head mounting for night driving, navigation, or handheld weapon usage. Additionally, for the dismounted application, an uncooled or short-wave infrared or forward-looking infrared camera will be added to the helmet-mounted assembly, providing thermal image insert to the image intensifier to enhance target detection performance and complement the image intensification performance (U.S. Army, 2003). This goggle will greatly enhance the vision of individual emergency responders in adverse visual conditions, especially vision restricted by lack of light or by smoke. It will also give them a much larger field of view than current systems do.	N
Long-Wave Micro-IR Sensor (Army S&T)	Develop miniature long-wave infrared thermal imagers based on advances in detectors, electronic components, and read-out integrated circuits. An intermediate result of this development effort was the Alpha camera, which went into production in 1999 as the world's first miniature thermal imager. The Omega camera, which went into full-scale production in 2002, improved on every significant aspect of its precursor.[c] The significant reduction in size and cost of the sensor and its capability to thermally image through smoke make it an ideal candidate to be added to the helmet of emergency personnel.	R
Urban Recon ACTD (OSD and National Geospatial-Intelligence Agency [NGA] S&T)	Provide a suite of terrestrial and airborne sensor and software capabilities enabling the warfighter to conduct effective urban terrain reconnaissance below the roofline, under the canopy, and within buildings. A user will be able to dynamically visualize a high-definition three-dimensional objective database in real time. The ACTD will develop applications for advanced urban decision aids and will leverage evolving technologies of geolocation and portable computing technology (U.S. Army, undated). With these technologies, emergency responders will have an enhanced capability for viewing the current situation within an urban environment, especially during and after a disaster.	R

continued

TABLE 4-4 Continued

Program	Description	Availability[a]
Networked Embedded Systems Technology (NEST) (Defense Advanced Research Projects Agency [DARPA] S&T)	Enable "fine-grain" fusion of physical and information processes. The quantitative target is to build dependable, real-time, distributed, embedded applications comprising 100 to 100,000 simple computing nodes. The nodes include physical and information system components coupled by sensors, actuators, and communications devices. NEST is an intelligent, Web-centric distribution and fusion of sensor information that will greatly enhance the situational awareness (friendly/enemy/civilian locations, sniper detection, and so on) of warfighters at lower echelons. It provides urban environment three-dimensional tracking of blue force personnel by allowing the warfighters to carry enough sensors (the size of a quarter) to "seed" a building while walking through it. The blue force will have radio frequency tags to stay connected to the NEST network (DARPA, undated). These technologies will allow for the continuous tracking of emergency responders in buildings, in subways, or in other situations where global positioning systems do not work.	F
Joint Biological Agent Identification and Agent Diagnostic System (OSD Joint Program Office [JPO] S&T)	Rapidly, reliably, and simultaneously identify multiple biological agents and pathogens. The ability to interface with electronic medical records/surveillance and early warning and reporting systems will occur in follow-on blocks (U.S. Army, 2003). This capability will allow emergency responders to detect the outbreak of a biological attack before it reaches epidemic proportions. It may also track day-to-day biological events such as the outbreak of flu epidemics.	N
Joint Biological Point Detection System (OSD JPO Acquisition)	Complete sensor suite with collector, automated assays, and detectors, as well as waste management, to identify 10 biological threat agents simultaneously in 20 minutes, as well as to collect liquid samples for confirmatory analysis. It is portable and can be installed in ships, vehicles, and fixed or semi-fixed sites. Eventually it is expected to identify up to 26 agents simultaneously. It can operate remotely at up to 5 kilometers and will interface with the Joint Warning and Reporting System (U.S. Army, 2003). This capability will allow emergency responders to detect the outbreak of a biological attack before it reaches epidemic proportions.	N

TABLE 4-4 Continued

Program	Description	Availability[a]
Joint Service Lightweight Integrated Suit Technology (OSD JPO Acquisition)	Protect against chemical or biological agents, produce a protective clothing ensemble that can be tailored to the diverse operational needs of the individual person and is compatible with existing and emerging protective clothing (U.S. Army, 2003). These suits will be of great value to emergency responders who are called to assist in a chemical or biological attack crisis situation.	N
Joint Service Lightweight Nuclear Biological Chemical Reconnaissance System (OSD JPO Acquisition)	Develop a system that consists of a base vehicle equipped with handheld, portable, and mounted, current and advanced nuclear, biological, and chemical identification equipment. The vehicle has collection, overpressure, navigation, meteorological data processing, internal and external communications, and surface sampler systems (U.S. Army, 2003). This equipment will provide emergency responders with a mobile, self-contained chemical/biological ISR capability.	N
Joint Service Lightweight Standoff Chemical Agent Detectors (OSD JPO Acquisition)	Identify chemically contaminated battlespaces and provide enhanced early warning. The detector is a passive, standoff, chemical detector for detection, identification, mapping, and reporting of nerve, blister, and blood agent vapors. This system can communicate with the Joint Early Warning and Reporting Network (U.S. Army, 2003). This will provide individual responders a standoff, static chemical/biological ISR capability.	N

[a]Availability: R, ready (TRL 8-9); N, near term (TRL 4-7); F, far term (TRL 1-3). See Appendix G in this report for descriptions of technology readiness levels (TRLs).

[b]One of the study committee members, Joseph P. Mackin, works for the company that supports this program—E-OIR Measurements, Inc., and so recused himself from specific discussion of this program.

[c]Stuart Horn, Science and Technology Division, Night Vision and Electronic Sensors Division, Communications-Electronic Research, Development and Engineering Center, "Uncooled Micro Sensors," briefing to the committee, Washington, D.C., August 26, 2003.

IR/Thermal Detector Technologies. The Army has had a leading role in developing IR and thermal detector technologies during the past five decades. The major technology investment has been in mercury cadmium telluride (HgCdTe). High-performance detector systems based on this technology are in use or under development. Other detector technologies include bolometers, quantum well infrared photodetectors, and Schottky barrier internal photoemission detectors. Germanium silicate (GeSi) heterostructure internal photoemission detectors, gallium antimony (GaSb) detectors, gallium nitride (GaN) detectors for ultaviolet

(UV) detection, and carbon nanotube arrays are other detectors that are in various stages of research and development for possible future applications and use. These were discussed in *Science and Technology for Army Homeland Security: Report 1* (NRC, 2003).

While emergency responders do not generally need the stringent capabilities of Army technology in this area, IR and thermal capability is of importance for firefighters as well as for perimeter defense and networked sensors. Again, while they may not have a need for solar blind optical detectors, UV semiconductor lasers and detectors currently being developed by the Defense Advanced Research Projects Agency (DARPA) solar blind UV detector programs as well as the Army's in-house and extramural research efforts will be useful for chemical and biological detection spectroscopy.

Nuclear, Radiological, and Explosive Threat Detection. The technical area relating to nuclear and radiological threat detection was discussed in Report 1 (NRC, 2003). The major conclusion of the committee was that for nuclear and radiological materials, the detection range of existing technologies and those under development was not long, and hence there were difficulties with standoff detection from any large distance. It was also pointed out in Report 1 that the lead responsibility for this area did not reside with the Army. However, the additional point was made that networked sensors and data fusion and management were critical Army areas of S&T investment and hence could have a strong impact in this area for emergency responders as well. For conventional explosives detection the situation is a little different. Report 1 discusses the different detection technologies and the challenges for these technologies owing to the low vapor pressure of more modern explosives. Again, Report 1 discusses the gains that could perhaps be made in looking at crosscutting technologies in addressing this problem. Furthermore some of the technologies for chemical and biological detection cross over very effectively into explosives detection as noted in the first report (NRC, 2003).

Chemical and Biological Agent Technologies. Chemical and biological agent detection technologies were also discussed in Report 1 (NRC, 2003). In the S&T arena, many sensors are in the preliminary research and development cycle. While the Army is the lead agency in this arena, the Joint Program Office, with a funding stream from the Office of the Secretary of Defense, is tasked with this responsibility. The vapor pressure of chemical agents is higher than that of explosives, but the acceptable exposure levels are lower, as was discussed in Report 1. Tables 2-2 and 2-3 in Report 1 list the different technologies that are in use and in various stages of research and development for detecting chemical and biological agents, respectively (NRC, 2003, pp. 50-53). The technologies relevant for this area overlap with those for detecting some other threats, especially conventional explosives. Hence, as discussed in the first report, crosscutting technologies may be very important here and may be candidates for collaborative research with the DHS (NRC, 2003).

Sensor Networking and Perimeter Sensors. The Army and other agencies have been pursuing the concept of networked sensors for several years, for use in both tactical situations and perimeter security. The general concept behind networked sensors is the ability to use disparate sensors, such as acoustic and seismic, non-imaging IR and laser, and visible and IR imaging, to develop a comprehensive situational awareness of an environment. Several schemes are available, with some currently focusing on the use of low-cost, low-power-consumption sensors such as seismic and acoustic and non-imaging IR and laser sensors to turn on the higher-cost, higher-power-consumption visible and IR sensors, generate an image or series of images, and then either send the images or processed information back to a central node or place them on a network for dissemination to users. Systems may include automatic alarms to indicate to the user if there is a disturbance in his or her area of operations. Several ongoing programs, described below, are taking this concept even farther, to include a moving infrastructure.

In August 2003, the Institute of Electrical and Electronics Engineers devoted a special issue to the topic Sensor Networks and Applications.[2] The issue contains nine papers, seven of which are invited. Eight of the nine papers describe work sponsored by DARPA. Many defense-related and homeland security applications are cited. The issue provides excellent coverage and is very up to date on the subject of sensor networks.

Synthetic Aperture Radar and Moving Target Indicator Technologies. The DOD has extensive programs in both synthetic aperture radar (SAR) and moving target indicator (MTI) technologies. The Army in particular has programs in small-scale, unmanned aerial vehicle-based systems for use at the tactical level that are also appropriate for homeland security purposes. The systems are expensive, however, and a trained force is required to maintain, operate, and interpret the data resulting from these radars. Thus, most nonmilitary crisis response organizations would not be sufficiently funded or staffed to have SAR/MTI radars as part of their organic equipment. SAR/MTI technologies are the type of capabilities that the DOD can bring to bear in support of emergency response during threats to homeland security. The interpreted output of the radars could then be integrated into the command-and-control network to give information to crisis managers. It is the product of the SAR/MTI, rather than the equipment itself, that is of value to homeland security crisis managers.

The primary benefit of a SAR is its all-weather capability. While a flying

---

[2]*Proceedings of the IEEE*, Volume 91, Issue 8, August 2003.

video camera is the cheapest and easiest way to see a swath of ground, it is of no value at night, in cloudy weather, or when obscured by smoke from fires at a crisis location. Currently available SARs can provide images with less than 1-foot resolution, which is adequate to give a picture of damage to facilities and the locations of vehicles and personnel (at the time of the image). Since the image requires time to process, only snapshots are available. The current SAR limitations are the latency in image processing and the lack of ability to see through foliage or inside structures. (Foliage penetration, or FOPEN, capability is about a decade away.)

The primary benefit of the MTI radar is its ability to track objects on the move. Military applications of this radar are to see which roads are being used for enemy attack or withdrawal; homeland security applications are to see which roads are blocked for access by responders or for the evacuation of personnel in dangerous areas such as in the path of a hurricane or a chemical attack cloud. The MTI radar also functions through adverse weather and obscurants.

Additional Department of Defense Assets for Consideration

The committee also calls attention to other programs and activities that, although not strictly within the C4ISR envelope of programs, may be of value to the emergency responder community. Several of these programs and activities are addressed below. Table 4-5 presents information on the relevant technologies related to other DOD assets, including a brief description or statement of purpose and availability assessment.

TABLE 4-5 Summary of Programs Relevant to Emergency Responders: Other Assets for Consideration

Program	Description	Availability[a]
Joint Virtual Battlespace (Army S&T)	Integrate common simulation environment and Army/joint simulations of varying fidelity with dynamic command-and-control and data flows that span the full battlefield spectrum from joint task force to entity level. This simulation environment will support engineering trade-off studies on the impact of information; information systems (sensors, communications, decision aids); and new tactics, techniques, and procedures (U.S. Army, 2003).	N
Effects of Weapons Simulation (Defense Threat Reduction Agency S&T)	Develop a chemical, biological, radiological, nuclear, and high explosive toolbox for simulation-based analysis and training. It will include (1) simulations for weapons effects, nuclear, biological, and chemical (NBC) environments, NBC defense, and command-and-control operations; and (2) high-fidelity, physics-based models and databases of	N

TABLE 4-5 Continued

Program	Description	Availability[a]
	targets, weapons, and after-strike effects that support the real-time/near-real-time viewing of the effects of weapons in a simulated environment. This simulation technology will also support the real-time visualization of the battlefield during exercises and live operations.[b]	
Flexible Asymmetric Simulation Toolkit (Defense Modeling and Simulation Office [DMSO] and U.S. Air Force S&T)	Develop a suite of warfighter-oriented tools for supporting decision making, mission planning, training, course-of-action analysis (to include tactics, techniques, and procedures), and mission rehearsal for operations other than war. It will provide capability to model movements of supplies, displaced personnel, supporting forces, and so on.[b]	N
Joint Conflict and Tactical Simulation (JCATS)—Laser Project (DMSO S&T)	Use laser-sensor technology to rapidly map complex terrain, including "rubble-ized" urban terrain, into a simulation (JCATS) for analysis.[c]	N
Dynamic Mission Readiness Training for C4ISR (Army and Air Force S&T)	Develop technology to improve planning, execution, and training by exploiting advanced training methods, mission rehearsal capabilities, and automated performance measurement and assessment technologies. Common training and mission rehearsal architectures will also be developed for distributed training, team training, and distributed team training including brief/debrief capabilities for pre-mission planning and post-mission assessment.[b]	N
Chemical and Biological Hazard Environment Prediction (Office of the Secretary of Defense S&T)	Develop an improved capability to predict the behavior of chemical and biological agents in the environment. It will address the physical and biological processes that affect chemical and biological agents after they have been released into the environment. These processes include transport, diffusion, deposition, evaporation, biological decay, and re-aerosolization.[b]	N

[a]Availability: R, ready (TRL 8-9); N, near term (TRL 4-7); F, far term (TRL 1-3). See Appendix G in this report for descriptions of technology readiness levels (TRLs).

[b]Selected information provided by the Office of the Director, Defense Research and Engineering, December 4, 2003.

[c]Personal communication between S.K. Numrich, Deputy Director for Technology, Defense Modeling and Simulation Office, and Albert A. Sciarretta, committee member, September 9, 2003.

Modeling and Simulation

The DOD has made a significant investment in modeling and simulation (M&S), which can be leveraged by the DHS. M&S tools can support the activities discussed below.

Planning. Before an event, M&S tools can be used to assess operational and support requirements, the long-term impact of a particular type of event (e.g., the release of a weapon of mass destruction), the displacement of personnel because of a crisis, and other related issues. Organizations such as the Army's Center for Army Analysis consistently use M&S tools to support combatant commands in planning their operational and logistical needs. The Army's Joint Virtual Battlespace Program can also be leveraged for these planning efforts. Planning tools for operations other than war (e.g., the Defense Modeling and Simulation Office/Army Flexible Asymmetric Simulation Toolkit) can be used for planning the movement of displaced personnel, the distribution of food and water, and support for personnel needs. Many of the Defense Threat Reduction Agency's tools can be used for assessing the effects of WMD.

Decision Support. Once an event has occurred, M&S tools similar to those mentioned above can be used for assisting the command element in making immediate decisions. M&S tools can enhance situation awareness—allowing a command element to better understand an environment that is sometimes defined by reams of data from many information and sensor systems. If linked to real-world sensor data and other relevant sources of information, these M&S tools can also support real-time predictions of chemical or biological cloud dispersion, traffic congestion, and so on.

Training. Training in the future will become ever more dependent on M&S because real-world and political considerations will make training more difficult. For example, large training exercises cannot be held in Washington, D.C., without taking many personnel away from their required jobs, without keeping large numbers of tourists out of the area, and without bringing media exposure to every success and failure of the event. Also, environmental considerations limit the types of training, as well as the availability of training areas. The costs of real-world training being higher, the ability to do some virtual training could lead to substantial cost savings. The inability of most emergency responders to attend training off-site will place even greater emphasis on M&S that can be used at responders' places of duty. Most importantly, and of increasing note, many ongoing security and operational missions are drastically reducing the time available for personnel to train. M&S can help hone skills in an embedded (or desktop) training environment, or through the use of the DOD's and the Army's Advanced Distributed Learning programs.

Testing

The Army, as well as other components of the DOD, has significant test range capacity and a system for testing equipment to ensure that it meets users' needs and manufacturers' claims. The testing includes both operational or performance testing and maintenance and logistics support testing. The Army has also developed, over many years, a testing methodology that allows the user to determine whether the equipment meets the full performance requirements promised—including reliability and maintainability. This could be of particular importance to local emergency responders, as the true cost of much of the equipment that they need is in the maintainability and repairs over a number of years.

Logistics

Critical to any successful system for use by either the military or emergency responders is a logistics and maintenance capability. Usually incorporated during system development, provision for logistics and maintenance supports the sustainability of an item of equipment for many years, ensuring that the equipment can be serviced, repaired as necessary, replaced, and disposed of at the end of its life cycle. It helps the user to determine the true life-cycle cost of equipment, what level of repairs are necessary (such as operator repair versus depot repair), and even disposal cost. It also helps assure the buyer that parts will be available for the foreseeable future and that equipment will not become obsolete owing to a future lack of such parts. The Army has many years of experience in this area, usually resident in its logistical and commodities centers.

Power Generation

Today's Army is making a significant effort to develop lighter, higher-energy-density hybrid power sources, chargers, and power management technologies for soldier systems; reformed logistic fuel components for fuel cells for vehicle-silent watch power; and fuel-efficient power generation and electronic control component technologies to provide for smaller, lighter, more-fuel-efficient mobile electric power generators. These new power systems would be of significant value to emergency responders. A report in preparation from the Board on Army Science and Technology concerning portable energy sources for the soldier sheds light on the actual progress being made in these areas (NRC, 2004).

SUMMARY

In this chapter the committee identifies some of the critical technical issues associated with a C4ISR system, highlighting several programs that may be of relevance to those developing such systems for use by emergency responders.

There is no attempt here to evaluate how well these programs are meeting their technical goals. However, where appropriate the committee expresses concerns about these technical objectives—particularly when they seem to be extremely difficult to achieve. The committee sincerely hopes that those who start down the path toward developing systems for emergency responders similar to those developed for the Army can at a minimum absorb some of the lessons learned from these prior endeavors and in some cases actually use the products of these programs.

REFERENCES

AITS-JPO (Advanced Information Technology Services–Joint Program Office). 2003. CINC 21. Available online at <http://www.les.disa.mil/cinc21.html>. Accessed November 19, 2003.

DARPA (Defense Advanced Research Projects Agency). Undated. Network Embedded Systems Technology. Available online at <http://dtsn.darpa.mil/ixo/programdetail.asp?progid=42>. Accessed October 22, 2003.

DOD (Department of Defense). 2003. Joint Blue Force Situational Awareness. Available online at <http://www.acq.osd.mil/actd/desctript.htm>. Accessed October 22, 2003.

Ericsson. 2001. Ericsson Response MiniGSM. Available online at <http://www.cs.berkeley.edu/~brewer/ict4b/Ericsson-miniGSM.PDF>. Accessed February 23, 2004.

JROC (Joint Requirements Oversight Council). 2003. Waveform Extract, Version A, Joint Tactical Radio System, Extract of JROC Approved Final with Waveform Table 4-2 and Annex E, Operational Requirements Document version 3.2, April 28. Washington, D.C: Joint Requirements Oversight Council.

NRC (National Research Council). 1997. The Evolution of Untethered Communications. Washington, D.C.: National Academy Press.

NRC. 2003. Science and Technology for Army Homeland Security: Report 1. Washington, D.C.: The National Academies Press.

NRC. 2004. Meeting the Energy Needs of Future Warriors. Washington, D.C.: The National Academies Press, in press.

Steinheider, J. 2003. Software-defined radio comes of age. February 11. Available online at <http://iwce-mrt.com/ar/radio_softwaredefined_radio_comes/>. Accessed December 16, 2003.

U.S. Army. Undated. Urban Recon ACTD, Airborne and Terrestrial 3-D Laser Imaging. Available online at <https://peoiewswebinfo.monmouth.army.mil/JPSD/UrbanRecon/Urban%20Recon%20Fact%20Sheet%20FINAL%2013MAY03.htm>. Accessed October 22, 2003.

U.S. Army. 2003. Weapon Systems 2003. Washington, D.C.: U.S. Government Printing Office.

5

Potential for Collaboration Between the Army and the Department of Homeland Security

The first four chapters lay the groundwork for this chapter by describing the Army's organizational structure, its Future Force, the needs of emergency responders, and planned and available technologies that can support homeland security. This chapter discusses ways to link science and technology (S&T) for the Future Force and emergency responder requirements. Additionally, it suggests how to facilitate collaboration between the Army and the Department of Homeland Security (DHS) for this purpose and it identifies issues associated with such collaboration.

POTENTIAL COLLABORATIVE EFFORTS TO ADDRESS UNMET NEEDS

The substantial overlap in the capabilities required by civilian emergency responders and by the Army confirms the potential of collaborative efforts between the Department of Defense (DOD), specifically the Army, and the DHS for transferring and adapting technologies and programs that underpin command, control, communications, computers, intelligence, surveillance, and reconnaissance (C4ISR) for the Army's Future Force. However, the committee cautions that not all local governments may be amenable to the idea of collaborative efforts, whether because of costs, perceived complexity of equipment, or simply a preference for remaining independent. Consideration might be given to the following possibility: whether it might be more reasonable and prudent initially to supply regional response teams—such as the National Guard's Civil Support Teams for Weapons of Mass Destruction—with appropriate C4ISR technology that can enable all of those responders to have compatible communications during an emergency or crisis event, rather than trying to field the same system for the entire first responder community.

Collaboration could take many possible forms. The efforts described here—leveraged collaboration and true joint development—represent two fruitful modes of operation that can assist the Army and the nation's emergency responders. It is analogous to the model established during the Cold War when Army and DOD science and technology provided many products that were very useful to the nation's commercial sector. There is also an interesting parallel between the possibilities presented here and the approach represented by the Foreign Military Sales Program or the Foreign Internal Defense Program long used by the military to encourage interoperability with allies and friends.

Leveraged Collaboration

Leveraged collaboration makes sense when one major user or developer is driving the process and another would like to leverage the ongoing efforts to reduce the cost or speed up its own product development. This is the most likely scenario in the near term for collaborative efforts between the Army and the emergency responder community, given the mature nature of most DOD research, development, testing and evaluation (RDT&E) infrastructure and the relative newness of the S&T organization within the DHS. It is also true, in general, that the DOD hardware requirements will often be different from and much more stringent than those for emergency responders (preparation for a warfighting environment can require qualities such as being air droppable, nuclear hardened, and so on) and therefore much more costly than emergency responders' budgets allow. However, technology developed for the warfighter can be adapted to the needs of the emergency responder community.

An example of a potential leveraged program *could* be the concept of the Joint Tactical Radio System (JTRS), Squad Level, discussed in Chapter 4. This software-programmable radio will support interoperability between the various DOD components in an integrated battlespace. While this type of radio would seemingly be very useful for the emergency responders in a local jurisdiction (such as the police, firefighters, National Guard, FBI, and so on), the DOD requirements with respect to the radio are well beyond what the local emergency responder needs and can afford. However, if the requirements can be adapted, with a corresponding reduction in costs, and if there is a sufficient market for such a concept for emergency responders, then there may be a potential for collaborative development.

Joint Development Collaboration

There may be cases in which a truly joint collaborative program between the DOD and the DHS is practical, particularly as the relationship between the two departments matures. In such a case, both departments would work together in developing technologies that would be of mutual benefit.

A Technological Bridge

Throughout its study, the committee examined the capabilities that are the foundations for the Future Force and those that could enable emergency responders in crisis situations. Bridging the two could help both communities leverage S&T to obtain capabilities needed to perform their respective missions. Table 5-1 indicates how technologies planned to meet Future Force requirements might address emergency responders' requirements in the following categories:

- Communications;
- Command, control, and computers;
- Intelligence, surveillance, and reconnaissance; and
- Other.

In the middle of the table—representing the area where the bridge must be established—technological opportunities for collaboration to meet the needs of both communities are identified as leveraged and joint activities. Table 5-1 is a tabular synopsis of the committee's assessment presented in the preceding chapters.

COLLABORATION ISSUES

In addition to technology transfer, the committee believes that there are other aspects to collaboration between the Army and the emergency responder community. These issues are addressed in the subsection below.

Systems Engineering

Transferring C4ISR technologies and systems to the DHS is not enough in itself to provide operationally suitable, supportable, and affordable C4ISR capabilities. Just as important as making technologies and systems available is that the Army also make its systems engineering expertise and experience available. This added benefit would enhance the DHS's current capability to execute a methodical systems engineering approach to satisfying its unique C4ISR needs, and in the longer term it can enhance compatibility between the Army and civilian emergency responders.

With this enhanced systems engineering capability, the DHS would have a running start and could pursue an integrated design approach to optimize the synergistic performance of a C4ISR system, or system of systems. Each component of each system, and each system within a system of systems, must be designed to function as part of a node (e.g., like a personal digital assistant) as

TABLE 5-1 Bridge Between Department of the Army/DOD Science and Technology for the Future Force and Emergency Responder Requirements

Aspect of C4ISR	Future Force Requirements	Leveraged Collaboration
Communications	Networked communications and data systems	Joint Tactical Radio System (JTRS) (Army Acquisition) JTRS Squad Level (Army S&T) Warfighter Information Network-Tactical (WIN-T) (Army Acquisition) Adaptive joint C4ISR node (Army S&T) Mobile network management (Army S&T)
Command, Control, and Computers	Act decisively	Smart Sensor Web (DUSD S&T) C3-on-the-move demonstration (Army S&T) Future command post technologies (Army S&T) Intelligent information technology (DARPA S&T) C2 in complex and urban terrain (Army S&T) Battle space terrain reasoning and awareness (Army S&T) Forecasting, planning, and resource allocation (USN, USAF, Army S&T) Geospatial information integration and generation (Army S&T) Agile Commander (Army S&T) Decision support systems for C2 (USN S&T) Homeland Security/DA ACTD (Army and DHS S&T) Joint Force Blue Force Tracking ACTD (OSD/DISA S&T) Knowledge fusion (Army S&T) FBCB2
Intelligence, Surveillance, and Reconnaissance	Know what the network knows	Smart Sensor Web (DUSD S&T) Land Warrior (Army Acquisition) Objective Force Warrior (Army S&T) Warfighter Physiological Monitoring System, part of Objective Force Warrior (Army S&T) Joint Intelligence, Surveillance, and Reconnaissance ACTD (OSD S&T) Network sensors for the Future Force (Army S&T) Advanced night vision goggles (Army S&T) Long-wave micro-IR sensors (Army S&T) Urban reconnaissance ACTD (OSD and NGA S&T) Network Embedded Systems Technology (DARPA S&T) UAVs/robotics Fusion-based knowledge for the Future Force Family of interoperable operational pictures

Joint Development Collaboration	Emergency Responder Requirements
Joint interoperable communications between DOD and local responders In-building communications and tracking global information grid	Networked communications and data systems
Decision-support tools and algorithms Information aggregation, fusion, and sorting Intelligence data dissemination to uncleared entities (soldiers or local responders) C4ISR interfaces for simulations	Informed event management
Joint development of chemical/biological/nuclear sensors Smart sensor networks for urban environments Low-cost, disposable, networked, multiphenomenology sensors Urban UAVs and robotics Space, airborne, and terrestrial sensors	Common operational picture

continued

TABLE 5-1 Continued

Aspect of C4ISR	Future Force Requirements	Leveraged Collaboration
Other	Other DOD assets	Joint Virtual Battlespace (Army S&T) Effects of Weapons Simulations (DTRA S&T) Flexible Asymmetric Simulation Toolkit (DMSO and USAF S&T) Joint Conflict and Tactical Simulation-Laser Project (DMSO S&T) Dynamic mission readiness training (Army and USAF S&T) Chemical and biological hazard environment prediction (USN S&T) Portable and mobile power (Army S&T)

NOTES: S&T, science and technology; DUSD (S&T), Deputy Undersecretary of Defense for Science and Technology; C3, command, control, and computers; DA, Department of the Army; DARPA, Defense Advanced Research Projects Agency; C2, command and control; USN, U.S. Navy; USAF, U.S. Air Force; ACTD, Advanced Concept Technology Demonstration; OSD/DISA, Office of the

well as being part of the network (network-centric environment). The overall performance of the C4ISR system will be further enhanced with consideration given to the human-system interface, system flexibility, reliability, maintainability, supportability, pre-planned product improvement, training, and safety.

The Army has the expertise, experience, and relevant industrial support to assist the DHS in designing a C4ISR architecture that could provide an effective and efficient path to developing C4ISR systems that are operationally robust at a more affordable life-cycle cost. A successful, cooperative Army-DHS C4ISR technology integration strategy must seek to meaningfully evolve appropriate C4ISR technologies, systems, and architectures (technical, systems, and operational) into a coherent DHS C4ISR capability. Without such a strategy, it will be difficult, if not impossible, for the DHS to properly integrate and transition C4ISR technologies.

A cooperative systems engineering approach will also help prevent the DHS from acquiring monolithic C4ISR systems that are not functionally compatible with military systems, that require access to huge databases, or that are combat-hardened with unnecessarily overdesigned, highly redundant hardware and software.

In some cases, the DHS may be better served by using the Army's systems engineering expertise to integrate commercial off-the-shelf technologies into a more affordable, uniquely designed C4ISR system. With the Army's assistance, more affordable, uniquely designed DHS systems can be created to interact,

Joint Development Collaboration	Emergency Responder Requirements
Virtual emergency exercises Plume and fire simulators	Other

Secretary of Defense/Defense Information Systems Agency; FBCB2, Force XXI Battle Command Brigade and Below; IR, infrared; NGA, National Geospatial-Intelligence Agency; UAVs, unmanned aerial vehicles; DTRA, Defense Threat Reduction Agency; DMSO, Defense Modeling and Simulation Office.

when needed, with fielded systems of military units employed as part of an emergency responder team.

To a considerable extent, the degree of interoperability will be dependent on the effectiveness of the software engineering effort. This effectiveness will in turn be dependent on the architecture of the hardware/software infrastructure, as well as the software development methodology and tool sets. A well-defined software architecture and software engineering environment are necessary precursors to efficient software design. The architecture must be defined at the basic processor/operating system level, at the interapplication communications infrastructure level, and at the additional support infrastructure level (e.g., intelligent information fusion agents).

With an Army-DHS systems engineering team approach, a rigorous configuration management plan can also be designed and executed to ensure that interoperability is maintained.

Finally, as a consideration, the DHS must recognize that to realize the anticipated benefits of network-centric C4ISR systems, emergency responder organizations will now need to have within their organizations, or at least have access to, network system administrators and other information technology professionals. Increased logistics support will also require repair persons, repair parts, and batteries. Despite requirements for increased resources, systems engineering can help minimize their impact.

Conclusion 5-1. The U.S. Army's proven experience in systems engineering can benefit the Department of Homeland Security's systems engineering efforts.

Recommendation 5-1. In addition to sharing command, control, communications, computers, intelligence, surveillance, and reconnaissance (C4ISR) technologies and systems, the U.S. Army should explore collaborative efforts to share pertinent systems engineering expertise with the Department of Homeland Security. These efforts should include the selection of applicable technologies for integration and systems engineering, such as the following:

- A systems architecture that provides an effective and efficient path to near-term systems acquisition and future technology insertion, and
- A technical architecture that ensures operational robustness and economic manufacturability.

Technology Transfer Coordination

The committee assesses and describes the Army's acquisition process in Chapter 1 and Appendix D. To facilitate technology transfer from the DOD to the DHS, the DHS needs to identify the capabilities required for emergency responders. The Army can assist in providing these needed capabilities by transferring already-existing technology and system solutions and/or those under development to the DHS.

Critical to the success of any type of collaboration is the ability to establish a meaningful dialogue between collaborating partners. There are several working groups that bring the Army together with the DHS, but it might be appropriate to establish a dedicated working group at the assistant secretary level to determine how these organizations can best work together. This approach would allow the Army to listen to the DHS's requirements and offer suggestions in an established forum. Such a working group meeting need not be a permanent event but can be tailored as work progresses.

Conclusion 5-2. A dedicated forum for the discussion of potential collaboration between the U.S. Army and the Department of Homeland Security could be a solid first step in establishing a mutually beneficial relationship.

Recommendation 5-2. The U.S. Army, working under the aegis of the Department of Defense, should establish a forum at the assistant secretariat level where it can meet with officials from the Department of Homeland Security to discuss how best to work together to encourage interoperability of communications and equipment and to take advantage of the economies of scale that might result.

The DHS Homeland Security Advanced Research Projects Agency (HSARPA) is responsible for extramural research and development (R&D) for the DHS. The agency will address a crosscutting portfolio of technologies and end users. Within this mission area, common technology areas should be identified that can be shared with the DOD. The DHS should identify the needs, roadmap, and requirements for mission success. One alternative to ensure optimal integration, sustainability, and accountability of federal technology investment is the establishment of mission-focused technology areas linked to capabilities- and performance-based requirements. By adopting such an approach, state and local communities will have enhanced opportunities for input to and refinement of the national technology investment.

Experimentation, Testing, and Review

The Army, as well as the rest of the DOD, has a very rigorous analysis process for evaluating new technologies and systems, as well as new or technology-driven operational or organizational concepts, doctrine, tactics, techniques, and procedures. This analysis process includes a combination of demonstrations, experiments, technical tests, and operational tests:

- *Demonstrations* normally provide limited views of technology or concept capabilities and do not include the rigorous collection of data. These demonstrations can be enhanced with some combination of models and simulations, breadboards in laboratories, representative technologies, and operationally ready systems in real-world environments. Demonstrations are normally designed to prove the utility of the best aspects of a technology or concept.
- *Experiments,* the next level up from demonstrations, include models and simulations, as well as well-defined scenarios, repetitive events, and rigorous data collection efforts.
- *Technical tests* focus on validating the technical performance of prototype (or limited production) systems or concepts. In addition, technical tests can be used to assess the manufacturer's stated performance for a given off-the-shelf item.
- Finally, *operational tests* take limited production and/or off-the-shelf technologies and systems and assess them in an operational environment with real users.

Technical and operational tests also involve very rigorous data collection efforts. Users can be included in demonstrations and experiments to speed up the assessment process and to allow user insights to be incorporated early in the development of a technology or system. Models and simulations are useful for supporting demonstrations, experiments, technical tests, and operational tests.

Technical and operational tests can be a combination of capabilities: live (e.g., pilots in real aircraft), virtual (e.g., pilots in flight simulators), or constructive (e.g., computer-generated aircraft). Models and simulations are normally used to reduce analysis costs, provide a scale-up capability, or allow analysis in an unsafe or environmentally unfriendly environment.

In support of available modeling and simulation and owing to the complexity of DOD systems and to the consequences of failure of these systems, the DOD maintains an extensive laboratory and testing and evaluation infrastructure to assist in evaluating the performance of these systems and the components that constitute them. This infrastructure is coupled to the DOD S&T programs through the oversight of the Director, Defense Research and Engineering (DDR&E). This program of modeling and simulation, experimentation, testing and evaluation, and review is in place to ensure that a system meets its specified functional and technical performance criteria and is operationally capable.

The systems that must be put in place to meet the objectives of the DHS may be similar in complexity to those developed by the DOD, and the consequences of any failure of these systems will be similarly grave. The DHS will likely find it necessary to institute a process to ensure that its systems meet specified functional and technical criteria and are operationally capable. Thus, an organization that can certify the validity of a manufacturer's claims for equipment performance must be available to the DHS.

To expedite the transfer of technology between the Army and the DHS, it would be helpful if the DHS instituted a modeling and simulation, experimentation, testing and evaluation, and review process similar to that used by the DOD. The Army could assist the DHS in identifying civilian and military subject-matter experts to serve as a testing and evaluation board, assessing the robustness and applicability of results to the end-user community. This community will be interdisciplinary, with a wide variety of backgrounds and many different requirements. The DHS could utilize the DOD infrastructure where it is appropriate and available. A vehicle for the coordination of these DHS and DOD activities and capabilities and needs should be put in place.

Conclusion 5-3. The systems that must be put in place to meet the objectives of the Department of Homeland Security will be similar in complexity to those developed by the Department of Defense, and the consequence of failure of those systems will be similarly grave.

Recommendation 5-3. The U.S. Army, through the Department of Defense, should offer to assist the Department of Homeland Security in developing critical capabilities, such as the following:

- A testing, evaluation, and review process;
- The spiral development process used by the Army; and
- Modeling and simulation.

Collaboration in Training Programs

A low-cost but robust, multidisciplinary, multilevel training and exercise program among all emergency responders and military response units will facilitate the integration of the multitudinous C4ISR technologies into an effective system-of-systems approach to emergency management. Additionally, such exercises will help bridge the civil-military cultural gap that can hinder interactions between the two quite different communities. The Army has established a comprehensive individual and unit training program that includes extensive joint exercises. The Army's National Simulation Center at Fort Leavenworth, Kansas, plays a significant role in achieving and maintaining a high level of training.

It is recognized that there is a program for training individual emergency responders as described in Chapter 3, but few multidisciplinary, multiechelon, all-hazards training and exercise programs are conducted on a continuing basis.[1] Not to be overlooked is the technical training and maintenance component required for the transfer of equipment and programs to emergency responders. Additionally, the personnel turnover within these organizations suggests the need for frequent joint exercises between emergency responders and the military.

There should be a mutual understanding of the respective training philosophies of civilian emergency responders and the military. Civilians need to be exposed to both Army doctrine and the equipment available in emergency situations, and vice versa. In some areas, this is being accomplished as a result of local Homeland Security Councils that include federal (FBI), county, city, and military installation personnel with emergency responder responsibilities.[2]

Conclusion 5-4. An immediate requirement exists for the coordination of comprehensive, multidisciplinary, multiechelon, all-hazards training and exercise programs between civilian emergency responders and the military.

Recommendation 5-4. The U.S. Army, through the Department of Defense, should offer to assist the Department of Homeland Security in coordinating all-hazards training and exercise programs for emergency responders and to make relevant Army training facilities available for these exercises.

---

[1]This shortcoming was reinforced during a meeting of the committee with the leadership of Columbus, Georgia. The police chief recognized that there was a capability for multiple sensor inputs but was desirous of a training/simulation exercise that would train emergency responders in how to make the inputs useful. Meeting of Michael F. Spigelmire on October 14, 2003, with the mayor, city manager, police chief, county sheriff, Fire and Emergency Medical Services chief, and director of prisons of Columbus/Muscogee County Consolidated Government, Georgia.

[2]July 14, 2003, meeting of Michael F. Spigelmire with Fort Benning, Georgia, Installation deputy commander and staff; and August 12, 2003, meeting with Eglin Air Force Base vice commander and staff.

Network-Centric Operations

The concept of network-centric operations (NCO) is as applicable to DHS's emergency response mission as network-centric warfare is to the Army's Future Force.[3] Both the Future Force and the nation's emergency responders will rely on a system-of-systems approach to *see*, *understand*, and *act* on situations. While network-centric warfare is a fairly well developed concept that enhances military capabilities and requirements, the concept of NCO for emergency responders is not as mature. Various sources of funds, different levels of technical capabilities, varied requirements, and the lack of an approved national operational framework for emergency response make NCO more difficult to implement in the near term. However, such a concept has merit when developing a long-term vision for the DHS. The ability to provide "the right information to the right people at the right time" (Cooper, 2003) could be enhanced by a concept of NCO that optimizes a system-of-systems approach to homeland security.

Developing such a vision and establishing a long-term roadmap to achieve it will help ensure the nation's best use of its limited resources. The seamless convergence of compatible equipment and the development of open standards for networking are essential components. By restricting federal grants to allow only the purchase of systems and capabilities that allow NCO, the leadership within the DHS can achieve the desired end state more rapidly.

The foundation of NCO is an integrated communications infrastructure that ties together key decision makers and emergency responders and allows sharing of critical information. This ubiquitous system could surge to meet time-sensitive demands for critical bandwidth in crises involving multiagency and multilocation events. Although such a national civilian integrated system does not now exist, the committee recognizes that efforts are under way to develop such a system.

Standardization Efforts

The U.S. military's process for defining required operational capabilities, known as Mission Essential Task Lists (METLs), define the tasks that a unit must perform, the conditions under which the tasks might take place, and the standards to which the tasks must be accomplished in order to complete a given mission. The tasks describe activities or objectives to be achieved; conditions define the environment under which the tasks have to be accomplished (e.g., weather conditions, personal protective equipment requirements); and standards provide measures of effectiveness for determining whether the tasks have been accomplished successfully.

[3]This subject has been discussed throughout the report, but the committee believes it to be of such significance to the overall focus of the report that it is again highlighted.

METLs are useful for directing training, measuring levels of readiness, defining requirements, and judging whether existing procedures and support systems are adequate. Hence, they offer a useful tool for identifying needs, clarifying requirements, and recognizing gaps. Currently, there is no universally accepted, analogous system at this level of specificity in the civilian emergency responder community. Many sets of standards for various activities related to responding to a terrorist attack do exist (Canada, 2003). However, there is not a set of nationally recognized, integrated tasks lists, particularly with regard to describing the tasks, conditions, and standards for local, state, and federal responses to catastrophic biological, chemical, nuclear, radiological, or explosive attacks.

Some have recommended establishing national preparedness standards—authoritative rules, principles, or measures to guide efforts in preparing for disasters. They argue that standards would improve coordination, identify gaps in capabilities, and promote higher levels of readiness (Canada, 2003). Adopting a system similar to the military's METL might achieve these ends.

In addition, emergency responder task lists would provide a standardized tool for comparing and identifying overlapping emergency responder and combat activities, helping to better identify common needs and opportunities for collaboration and technology investment. As the committee previously pointed out, product standards and conformity testing are extremely important. In the Army's case, this evaluation includes both testing in the development laboratories and operational testing in the field by soldiers. The responder community, especially in the smaller jurisdictions, lacks access to the full process. Only when there are well-crafted standards for equipment used by responders, as well as testing laboratories that are regularly subject to a formal laboratory certification procedure, will responders be assured that they are purchasing items that meet their needs.

Conclusion 5-5. Emergency responders lack a standardized means to define the capabilities required to respond to a terrorist attack.

Recommendation 5-5. The U.S. Army, primarily through the local Army National Guard structure, should assist emergency responders by working with the Department of Homeland Security to begin to develop a process for defining a set of tasks similar to the process underlying the Army's Mission Essential Task List.

Conclusion 5-6. Common product standards and conformity testing are necessary to ensure interoperability between technology materiel of the Department of Defense and equipment used by emergency responders.

Recommendation 5-6. The Department of Defense and the Department of Homeland Security should jointly develop analytical tools for determining common equipment needs based on common group task analysis so as to establish common product standards for emergency responder technology materiel.

SUMMARY

The requirement for C4ISR is ubiquitous, whether for the Army's Future Force or for today's emergency responder. The committee is convinced that quick action on the part of the Army can provide beneficial C4ISR solutions to the Department of Homeland Security that will ensure a high level of interoperability between the emergency responders and the Army should our nation be forced to respond again to a catastrophic event on U.S. soil.

REFERENCES

Canada, B. 2003. Homeland Security: Standards for State and Local Preparedness, May 12. Available online at <http://public.ansi.org/ansionline/Documents/Standards%20Activities/Homeland%20Security%20Standards%20Panel/RL31680.pdf>. Accessed November 20, 2003.

Cooper, S. 2003. Presentation by Steve Cooper, Chief Information Officer, Department of Homeland Security, to the National Institute for Urban Search and Rescue Conference, January 13, 2003, San Diego, Calif.

6

Complete List of Findings, Conclusions, and Recommendations

OVERARCHING RECOMMENDATION

Recommendation. The Department of the Army, in coordination with the Department of Defense, should carry out the following:

- Work with the senior leadership in the Department of Homeland Security (DHS) to put in place and *to institutionalize a process for collaboration and sharing* between the Army and the DHS;
- Assist the DHS in *establishing the research, development, testing, and evaluation infrastructure* (i.e., an acquisition process, systems engineering discipline, modeling and simulation technologies, and testing and evaluation facilities) *to support the emergency responder community*;
- Work with the DHS *to find common areas of science and technology collaboration*, starting with the Future Force technologies identified in this report. Central to this effort will be the development of a framework or architecture to enable the integration of these technologies into an effective system of systems; and
- Work with the DHS *to establish processes for joint*[1] *operations*, including joint training and exercises, shared standards, and interoperable systems.

[1]Joint in this application means between civilian and military.

FROM CHAPTER 1, "INTRODUCTION"

Finding 1-1. Although a number of informal mechanisms exist, no coherent planning paradigm for the interface between the military and the emergency responders currently exists, and although a national operational concept for emergency response is being developed, it is not yet a comprehensive framework that pulls together the efforts of federal, state, and local responders.

Finding 1-2. The U.S. Army has developed a number of capabilities that could be used by emergency responders:

- Relevant technologies from the Army science and technology base;
- C4ISR systems that have been developed and deployed by the Army;
- An acquisition system, similar to the Army's spiral development process, that encompasses identifying needs, funding the required technology, and developing fieldable products;
- A testing and certification process for new equipment;
- Training programs;
- A network-centric operations approach;
- Exercises (and supporting facilities);
- Modeling and simulation capabilities; and
- A process for the development and assessment of doctrine.

FROM CHAPTER 2, "CAPABILITIES FOR THE ARMY'S FUTURE FORCE"

Finding 2-1. The network-centric concept is the foundation of the Army's Future Force.

Conclusion 2-1. The U.S. Army possesses a large and varied number of Future Force science and technology programs that, with proper coordination, could be made available to the Department of Homeland Security; however, there is currently no planning process to identify which could be shared or how to do so.

Recommendation 2-1. The U.S. Army, through the Department of Defense, should work with the Department of Homeland Security to analyze and determine, among other items, appropriate planning processes necessary to determine which Future Force science and technology programs should be shared, and how best to go about doing this.

FROM CHAPTER 3, "CAPABILITIES FOR EMERGENCY RESPONDERS"

Conclusion 3-1. Once fully established, the national requirements for command, control, communications, computers, intelligence, surveillance, and reconnaissance (C4ISR) technologies to support emergency responders will be substantial and sustainable and could create a significant market.

Conclusion 3-2. Individual emergency responder C4ISR systems need to be linked and integrated into a national operational framework.

Recommendation 3-2. The U.S. Army, through the Department of Defense, should offer to assist the Department of Homeland Security in developing a concept of operations for a national operational framework, to include the appropriate architectures and enabling technologies for C4ISR.

FROM CHAPTER 4, "DEFENSE TECHNOLOGIES FOR HOMELAND SECURITY"

Conclusion 4-1. The U.S. Army has developed a significant number of C4ISR technologies for the Future Force that appear to have direct applicability to the emergency responder community.

Recommendation 4-1. The U.S. Army and the Department of Homeland Security should evaluate the systems described in Chapter 4 of this report for their potential to support interagency collaboration.

FROM CHAPTER 5, "POTENTIAL FOR COLLABORATION BETWEEN THE ARMY AND THE DEPARTMENT OF HOMELAND SECURITY"

Conclusion 5-1. The U.S. Army's proven experience in systems engineering can benefit the Department of Homeland Security's systems engineering efforts.

Recommendation 5-1. In addition to sharing command, control, communications, computers, intelligence, surveillance, and reconnaissance (C4ISR) technologies and systems, the U.S. Army should explore collaborative efforts to share pertinent systems engineering expertise with the Department of

Homeland Security. These efforts should include the selection of applicable technologies for integration and systems engineering, such as the following:

- A systems architecture that provides an effective and efficient path to near-term systems acquisition and future technology insertion, and
- A technical architecture that ensures operational robustness and economic manufacturability.

Conclusion 5-2. A dedicated forum for the discussion of potential collaboration between the U.S. Army and the Department of Homeland Security could be a solid first step in establishing a mutually beneficial relationship.

Recommendation 5-2. The U.S. Army, working under the aegis of the Department of Defense, should establish a forum at the assistant secretariat level where it can meet with officials from the Department of Homeland Security to discuss how best to work together to encourage interoperability of communications and equipment and to take advantage of the economies of scale that might result.

Conclusion 5-3. The systems that must be put in place to meet the objectives of the Department of Homeland Security will be similar in complexity to those developed by the Department of Defense, and the consequence of failure of those systems will be similarly grave.

Recommendation 5-3. The U.S. Army, through the Department of Defense, should offer to assist the Department of Homeland Security in developing critical capabilities, such as the following:

- A testing, evaluation, and review process;
- The spiral development process used by the Army; and
- Modeling and simulation.

Conclusion 5-4. An immediate requirement exists for the coordination of comprehensive, multidisciplinary, multiechelon, all-hazards training and exercise programs between civilian emergency responders and the military.

Recommendation 5-4. The U.S. Army, through the Department of Defense, should offer to assist the Department of Homeland Security in coordinating all-hazards training and exercise programs for emergency responders and to make relevant Army training facilities available for these exercises.

Conclusion 5-5. Emergency responders lack a standardized means to define the capabilities required to respond to a terrorist attack.

Recommendation 5-5. The U.S. Army, primarily through the local Army National Guard structure, should assist emergency responders by working with the Department of Homeland Security to begin to develop a process for defining a set of tasks similar to the process underlying the Army's Mission Essential Task List.

Conclusion 5-6. Common product standards and conformity testing are necessary to ensure interoperability between technology materiel of the Department of Defense and equipment used by emergency responders.

Recommendation 5-6. The Department of Defense and the Department of Homeland Security should jointly develop analytical tools for determining common equipment needs based on common group task analysis so as to establish common product standards for emergency responder technology materiel.

Appendixes

Appendix A

Biographical Sketches of Committee Members

John W. Lyons, NAE, *Chair,* consultant and retired director of the Army Research Laboratory (ARL), is a Ph.D. physical chemist. He served in research and development positions with the Monsanto Company for 18 years. In 1973 he joined the Commerce Department's National Bureau of Standards (NBS). At NBS, he was the first director of the Center for Fire Research. In 1990 Dr. Lyons was appointed by President George H.W. Bush to be the ninth director of NBS, by that time renamed the National Institute of Standards and Technology (NIST). In September 1993, he was appointed the first permanent director of ARL. At ARL, Dr. Lyons managed a broad array of science and technology programs. He has served on many boards and commissions, inter alia, the Federal Advisory Commission on Consolidation and Conversion of Defense Research and Development Laboratories. He currently serves on two boards of visitors at the University of Maryland. He is a member of the National Research Council's Board on Army Science and Technology, as well as a member of a congressionally chartered committee at the National Defense University to study the potential effectiveness of the DOD laboratories in the transformed military of the future. Dr. Lyons was elected to the National Academy of Engineering in 1985. He is a fellow of the American Association for the Advancement of Science and of the Washington Academy of Science and is a member of the American Chemical Society and of Sigma Xi.

Dennis J. Reimer, *Vice Chair,* is director of the National Memorial Institute for the Prevention of Terrorism, Oklahoma City. The institute is dedicated to preventing, reducing, and mitigating the effects of terrorism, with particular emphasis on the role of first responders. A retired U.S. Army general, he was most recently the 33rd Chief of Staff of the Army. Prior to his term as Chief of Staff,

General Reimer commanded all Army forces (except Special Forces) assigned to the continental United States. He holds a B.S. from the U.S. Military Academy at West Point and an M.S. from Shippensburg State College.

Duane A. Adams is currently vice provost for research at Carnegie Mellon University. He holds a B.A. in mathematics from the University of Montana and a Ph.D. in computer science from Stanford University. Previously, Dr. Adams served at high levels of government in programs relating to advanced computer research, including service in the Office of the Assistant Secretary of Defense (Program Analysis and Evaluation). Dr. Adams has also been a member of the Air Force Scientific Advisory Board and a member, vice chair, and chair of the Army Science Board. In addition, he has served as a member of several National Research Council committees, including the Committee on Air Force Base Level Automation and the Committee to Study International Developments in Computer Science and Technology.

Henry L. Bertoni is head of the Department of Electrical and Computer Engineering at Polytechnic University in New York. He received a Ph.D. in electrophysics in 1967 from Polytechnic Institute of Brooklyn (now Polytechnic University). Since the mid-1980s, Dr. Bertoni has led a group in the study of ultrahigh-frequency propagation in urban environments. He and his associates were the first to understand the mechanisms governing average signal strength for elevated base station antennas of cellular mobile radio. The results of these advances are the basis for the COST-231 model used throughout the world for installation of a 1900-MHz global system for mobile communications (GSM) and personal communications service (PCS) systems. Dr. Bertoni's group has also studied characteristics of the indoor radio channel both theoretically and experimentally. These studies have led to ray tracing codes predicting indoor propagation. Dr. Bertoni has been widely published and is a fellow of the Institute of Electrical and Electronics Engineers (IEEE) and a member of the International Scientific Radio Union and the Radio Club of America. He has served as chair of the Technical Committee on Personal Communications of the IEEE Communications Society, and as chair of the Hoover Medal Board of Award.

James J. Carafano is the senior research fellow for defense and homeland security in the Kathryn and Shelby Cullom Davis Institute for International Studies at the Heritage Foundation in Washington, D.C. Dr. Carafano joined the foundation after serving as a senior fellow at the Center for Strategic and Budgetary Assessments, a Washington policy institute dedicated to defense issues. Before that, he served for 25 years in the Army, rising to the rank of lieutenant colonel. During his service, Dr. Carafano served in Europe, Korea, and the United States and was a special assistant to the Army Chief of Staff, the service's highest-ranking officer. Before retiring, he was executive editor of the *Joint Force*

Quarterly, the Department of Defense's principal professional military journal. Dr. Carafano also taught military history at the U.S. Military Academy West Point and the U.S. Army Field Artillery School and served as the director of military studies at the Army's Center of Military History. He continues to teach as an adjunct professor at Georgetown University and at the U.S. Naval War College. He is the author of two books: *Waltzing into the Cold War* (published in 2002 by Texas A&M University) and *After D-Day: Operation Cobra and the Normandy Breakout*, a Military Book Club selection (published in 2000 by Lynne Rienner). A graduate of West Point, Dr. Carafano also has a doctorate from Georgetown University and a master's degree in strategy from the U.S. Army War College.

George M. Clark, as president and cofounder of Radiance Technologies, has led the development of the company to its current size (projected sales in 2003 of $15.5 million, and 90 employees). Dr. Clark managed the development of the Small Arms Tactical Recognition Equipment (STARE) System that supports ground forces by detecting, classifying, and locating small-caliber weapons in real time. He is currently leading the Overwatch Advanced Concept Technology Demonstration, a direct follow-on of STARE. He led the Radiance (STD) Program Analysis and Concept Engineering (SPACE) Program for the Space Technology Directorate of the Army's Missile and Space Technology Center. In that capacity, Dr. Clark provided technical expertise and leadership in the areas of system engineering, system design, test planning, and programmatic support for STD programs, including the Battlefield Ordnance Awareness Program, the Radar Power Technology Program, the Overhead Sensor Program, and other space technology activities. Dr. Clark holds a Ph.D. from the Georgia Institute of Technology.

Timothy Coffey holds a Ph.D. in physics from the University of Michigan. Dr. Coffey joined the Naval Research Laboratory (NRL) in 1971 as head of the Plasma Dynamics Branch in the Plasma Physics Division. In this position, he directed research in the simulation of plasma instabilities, the development of multidimensional fluid and magnetohydrodynamic codes, and the development of computer codes for treating chemically reactive flows. In 1975, he was named superintendent, Plasma Physics Division; he was appointed associate director of research for general science and technology on January 1, 1980. On November 28, 1982, he was named director of research. Dr. Coffey retired from the NRL in October 2001 and joined the University of Maryland. He has contributed to the theory of nonlinear oscillations and has played a major role in the national program on high-altitude nuclear effects. The author or coauthor of more than 70 publications and reports, Dr. Coffey has made several fundamental contributions to the theory of electron beam/plasma interaction and to the understanding of plasma processes in Earth's ionosphere. Dr. Coffey is a fellow of the American

Physical Society, the Washington Academy of Science, and the Franklin Institute, and a member of the American Physical Society, the American Association for the Advancement of Science, and Sigma Xi. He was awarded the Delmer S. Fahrney Medal and the Department of Defense's Distinguished Service Medal in 1991, and in August 2000 he was awarded the Navy's prestigious Captain Robert Dexter Conrad Award. Upon his retirement from the Naval Research Laboratory, he was awarded the NRL Lifetime Achievement Award.

Anthony C. DiRienzo is currently the executive vice president and chief technology officer of COLSA Corporation, located in Huntsville, Alabama. Dr. DiRienzo oversees the operations of 75 different programs of various government contracts, including radar hardware-in-the-loop development, large-scale computing network development, advanced signal processing algorithms, intelligence program support, acquisition and force management support, missile defense test and evaluation, integrated system testbed development, complex system integration programs, and software independent validation and verification. Previously, from 1995 to 1998, he directed the joint Army-Ballistic Missile Defense Organization $150 million contract to construct the national missile defense radar prototype located at the Reagan Test Site in the Pacific. Additionally, his professional activities have included directing the Army/Marine Corps Firefinder field artillery counterbattery radar program and serving as a staff officer in the Army Secretariat with responsibility for wide-ranging classified vulnerability assessment programs for Army weapon systems. He holds an M.A. from Georgetown University in international security and an M.S. in nuclear physics and a Ph.D. in plasma physics from the Massachusetts Institute of Technology.

Mitra Dutta currently serves as professor and head of electrical and computer engineering, as well as adjunct professor of physics, at the University of Illinois at Chicago. She received her M.S. and Ph.D. degrees from the University of Cincinnati. She has had appointments at the College of Arts and Sciences at Kingston, Jamaica, West Indies; postdoctoral appointments at Purdue University and City College of New York; and adjunct professor appointments at Rutgers University, the University of Maryland, North Carolina State University, and the University of North Carolina at Chapel Hill, as well as at Brookhaven National Laboratory. She worked for 15 years at the U.S. Army Research Laboratory in various capacities, and prior to joining the faculty of the University of Illinois, Dr. Dutta served in a senior executive service position in the Army Research Office (ARO), now a component of the U.S. Army Research Laboratory. She has authored or coauthored more than 370 publications and presentations and holds 29 U.S. and Canadian patents. She is a fellow of the Institute of Electrical and Electronics Engineers (IEEE), the American Association for the Advancement of Science, and the Optical Society of America. Dr. Dutta was the recipient of the U.S. Army Research and Development Achievement Awards in 1990, 1992, and

1995, the Harold Jacobs Award in 1991, the Paul A. Siple Award in 1994, the IEEE Harry Diamond Award in 2000, and the National Award for Achievement from the Society of Women Engineers in 2003.

Frederick L. Frostic is currently a principal with Booz Allen Hamilton. Prior to joining Booz Allen, he served as Deputy Assistant Secretary of Defense for Requirements and Plans, preparing the Defense Planning Guidance, supervising the Department OF Defense's response to the congressionally mandated Commission on Roles and Missions, and conducting crisis planning, reviews of plans, and force structure analysis. Recently, he was the project manager of a team providing research to the U.S. Commission on National Security/21st Century (Hart-Rudman Commission). In this effort, his team wrote the implementation plan for the commission's recommendations on homeland security. Additionally he was the project manager for providing research support to the Presidential Commission on Critical Infrastructure Protection. Mr. Frostic, a graduate of the Air Force Academy, earned an M.S. in engineering from the University of Michigan in 1971.

C. William Gear, a member of the National Academy of Engineering, is president emeritus of the NEC Research Institute. Prior to joining NEC, he was head of the Department of Computer Science and professor of computer science and applied mathematics at the University of Illinois at Urbana-Champaign. His research expertise is in numerical analysis and computational software. Dr. Gear is a fellow of the American Academy of Arts and Sciences, the Institute of Electrical and Electronics Engineers, the American Association for the Advancement of Science, and the Association for Computing Machinery (ACM). He served as president of the Society for Industrial and Applied Mathematics and was the recipient of the ACM SIGNUM George E. Forsythe Memorial Award and Fulbright and Johnson Foundation fellowships.

James R. Klugh is currently the technical director and vice president for information technology for Dimensions International, Inc. A retired Army major general, his last military position was as assistant deputy chief of staff for logistics at Headquarters, Department of the Army. He is a nationally recognized leader with extensive experience in command, control, communications, computers, intelligence, surveillance, and reconnaissance. A graduate of South Carolina State University with a B.S. in chemistry and mathematics, Mr. Klugh also has an M.S. in administration and management from Shippensburg State College in Pennsylvania. He served as director of the Department of Defense's chemical and biological research, development, and defense programs and has also developed plans and managed activities in response to chemical, biological, and nuclear incidents. Mr. Klugh established a joint total asset visibility program for tracking supply support to all armed forces, including the National Guard and the Reserves.

This program included the use of best technology solutions in radio frequency, satellite tracking, and automatic identification equipment. The global technical architecture of tracking and reporting devices established the foundation for in-transit visibility of personnel, equipment, and supplies across the Department of Defense.

Joseph P. Mackin is currently president of E-OIR Measurements, Inc., a sensor applications company in Virginia. He has an extensive background in sensors, having served in many Department of Defense sensor development and acquisition assignments. He was the project officer for the Air Force office developing sensors for high-valued assets such as nuclear weapons, deputy division director of the Laser Division at the U.S. Army Night Vision and Electronic Sensors Directorate, product manager for the Army's second-generation FLIR (thermal imager) for the Abrams Tank and Bradley Fighting Vehicle, and the director of special programs on the staff of the Army acquisition executive. Since retiring from the Army, and prior to accepting his current position, he worked at Massachusetts Institute of Technology (MIT) Lincoln Laboratory as an assistant group leader in the Sensors Applications Group, where he was the technical lead for the Deputy Undersecretary of Defense for Science and Technology's Smart Sensor Web program. His education includes a B.S. from the U.S. Military Academy at West Point, an M.S. in physics from the Naval Postgraduate School, and a Ph.D. in physics from the Massachusetts Institute of Technology. He is also a graduate of the Defense Systems Management College.

Louis C. Marquet, received his B.S. degree from the Carnegie Institute of Technology (now Carnegie Mellon University) and an M.S. and a Ph.D. in physics from the University of California at Berkeley. Now retired from the Civil Service, he serves as a private consultant. His previous position was director of the Army's Communications and Electronics Command Research, Development, and Engineering Center at Fort Monmouth, New Jersey. Prior to this assignment, Dr. Marquet held a number of senior government positions, including director of the Army's Night Vision and Electronic Sensors Directorate, assistant deputy undersecretary for technology in the Office of the Secretary of Defense, deputy for technology at the Strategic Defense Initiative Organization (now Missile Defense Agency), and director of the Directed Energy Office at the Defense Advanced Research Projects Agency. Additional positions that Dr. Marquet has held include vice president at the Nichols Research Corporation and the Atlantic Aerospace Electronics Corporation, associate head of the Optics Division at the MIT Lincoln Laboratory, assistant professor of physics and astronomy at the University of Arizona, Tucson, and assignment on active duty with the U.S. Army (Signal Corps). Dr. Marquet has received numerous official awards and recognition, including the Office of the Secretary of Defense Meritorious Civilian Service Award (twice), the Presidential Rank Award for Meritorious Executives in 1999, the Senior Executive Association Professional Development League 1998 Executive Achievement Award, and the ADPA 1987 Strategic Defense Award. Most recently, he was awarded the AFCEA Benjamin H. Oliver Gold Medal for Engineering for 2000.

Lois C. McCoy is president of the National Institute for Urban Search and Rescue (NIUSR), an organization that established the first urban rescue teams in California in 1981. These teams are now designated as Federal Emergency Management Agency Task Forces; they responded to the disasters at the World Trade Center and the Pentagon on September 11, 2001. Ms. McCoy was one of the founding members of the institute in 1977, serving first as its CEO and then becoming president in 1981. NIUSR is a principal national center for the application of new technology toward lifesaving improvements from the field up through the policy level. Its core cadre of 250 is chosen from across the diverse fields of first responders, military institutions, government at all levels, industry, and academia. Previously, Ms. McCoy had been in the field of lifesaving for more than 30 years, beginning as a founding member of the prestigious San Diego Mountain Rescue Team. She came up through the ranks of the emergency management field, through mountain and desert rescue, as an emergency medical technician, communicator, operation leader, government liaison, military liaison, urban task force developer, and county government emergency manager and coordinator. Ms. McCoy currently leads the joint executive board of NIUSR.

Chandra Kumar N. Patel, a member of the National Academy of Engineering and the National Academy of Sciences, is chief executive officer and chairman of the board of Pranalytica, Inc., and a professor of physics and former vice chancellor of research at the University of California at Los Angeles. Until 1993, Dr. Patel served as executive director of the Research, Materials Science, Engineering, and Academic Affairs Division at AT&T Bell Laboratories. Dr. Patel has an extensive background in several fields, including materials, lasers, and electro-optical devices. During his career at AT&T, which began in 1961, he made numerous seminal contributions in several fields, including gas lasers, nonlinear optics, molecular spectroscopy, pollution detection, and laser surgery. Dr. Patel has served on numerous government and scientific advisory boards, and he is a past president of Sigma Xi and the American Physical Society. In addition, he has received numerous honors, including the National Medal of Science for his invention of the carbon dioxide laser.

Albert A. Sciarretta is president of CNS Technologies, Inc., a company that consults on research and development, experimentation, modeling and simulation, management, and assessment of advanced information, sensor, and test technologies. He recently served as experiment director of the Department of Defense's Smart Sensor Web effort and as director of a demonstration of an integrated live-virtual-constructive simulation-based joint urban operations training environment. His current primary efforts include demonstrating networked sensor-information systems, assisting in the development of command and control (C2) systems for urban operations, assessing advanced information and test technologies, and identifying performance metrics for the Army's Future Force Warrior and associated small-unit C2 systems. Mr. Sciarretta is a retired Army officer. He has a B.S. degree from the U.S. Military Academy and dual M.S. degrees in mechanical engineering and operations research from Stanford University. He previously served as a member of the National Research Council's (NRC's) Committees on Review of the Department of Defense Air and Space Systems Science and Technology Program, Army Unmanned Ground Vehicle Technologies, and Advanced Energetic Materials.

Annette L. Sobel is a distinguished member of the technical staff of Sandia National Laboratories and a systems analyst with 13 years of experience in advanced technology development and unconventional threat analysis. She is currently serving as the director of homeland security for the state of New Mexico. Her research interests focus on applications of biotechnology and information technologies in support of chemical-biological countermeasures and in the field of human factors/systems engineering (e.g., critical decision making under stress) domains. Her work has emphasized information analysis, advanced systems for mission rehearsal and training, human performance enhancements, and technology transition to field operational environments. She is a Brigadier General in the Air National Guard and the J2, director of intelligence for the National Guard Bureau, and previously special assistant for weapons of mass destruction and civil support. She has 11 years of military command experience that includes combat and chemical-biological warfare medical response unit commands. Dr. Sobel earned an M.D. at Case Western Reserve University, with a specialization in family medicine at Duke University Medical Center. She has an M.S. in aerospace medicine with an emphasis on human factors engineering from Wright State University. Currently, she is a member of the Defense Intelligence Agency's advisory board.

Michael F. Spigelmire is a consultant on crisis response, consequence management, and force protection. A retired U.S. Army lieutenant general, he has had a military career with a unique blend of conventional and special operations assignments. General Spigelmire commanded the U.S. Army's Special Operations Command and then the VII Corps in Germany. Upon retirement, he assumed the position of deputy director of operations for the Atlanta Committee for the Olympic Games. This brought him into close contact with municipal, state, and federal officials. General Spigelmire holds an M.A. in international relations from Georgetown University. Additionally, he has completed studies at the U.S. Army Command and Staff College and the U.S. Army War College. General

Spigelmire is currently the senior mentor for the Terrorist Response Senior Seminar, sponsored by the Joint Special Operations University and the Air Force Special Operations School, Hurlburt Field, Florida.

Leo Young, a member of the National Academy of Engineering, retired as director for research and laboratory management in the Office of the Director for Defense Research and Engineering in the Office of the Secretary of Defense in 1994 and consulted almost full-time for that office until 2002. Since 1994, he has been on the Technology Advisory Board of Filtronic, an international company headquartered in the United Kingdom with research and manufacturing facilities in the United States. Dr. Young has held senior positions at the Naval Research Laboratory, the Stanford Research Institute, and the Westinghouse Electric Corporation. Dr. Young holds honors degrees in physics and in mathematics from Cambridge University and a doctor of engineering degree from Johns Hopkins University, which also awarded him the honorary degree of Doctor of Humane Letters, as well as the Woodrow Wilson Award for Distinguished Government Service. He has served on the Advisory Board of the Johns Hopkins University's Whiting School of Engineering and is currently chairman of the External Advisory Committee to the Department of Electrical and Computer Engineering. Dr. Young has authored, coauthored, or edited 14 books and more than 100 papers, and he holds 20 patents. He is a fellow and past president of the Institute of Electrical and Electronics Engineers, a fellow of the American Association for the Advancement of Science, and a fellow of the Royal Academy of Engineering of the United Kingdom. Dr. Young has also served on several NAE committees, including the National Academies' Government-University-Industry Research Roundtable.

Appendix B

Committee Meetings

FIRST MEETING

July 21-22, 2003
Washington, D.C.

Meeting objectives: National Research Council introduction; complete administrative actions, including committee introductions, composition/balance/bias discussions for committee members, and committee and report procedures; discuss statement of task with sponsor, discuss draft report outline, project plan, and report realization; make writing assignments; confirm objectives, location, and dates for the next two committee meetings.

Presenters

Sponsor Discussion Time
John Parmentola, Director of Research and Laboratory Management

C4ISR for the Washington, D.C., Fire Department
Michael Sellitto, Deputy Chief for Special Operations, and Peter LaPorte, Director of Emergency Operations

Army C4ISR Technology for Homeland Defense
Larry L. Fillian, Director, Command and Control Directorate, Communications and Electronics Research, Development and Engineering Center

C4ISR Technology for the Objective Force
Larry L. Fillian, Director, Command and Control Directorate, Communications and Electronics Research, Development and Engineering Center

C4ISR Requirements for the Nation's First Responders
Guy W. Beakley, Vice President of C4ISR, Hicks & Associates, Inc.

SECOND MEETING

August 25-26, 2003
Washington, D.C.

Meeting objectives: Complete composition/balance/bias discussions for committee members; examine joint and service doctrine describing the mission of the Army in HLS; examine C4ISR requirements for civilian emergency responders; preview C4ISR technologies that may have collaboration potential; discuss project plan and report realization; discuss concept draft, make additional writing assignments; confirm objectives, location, and dates for the next two committee meetings.

Presenters

C4ISR Requirements for the Army's Objective Force
Lieutenant General Johnny M. Riggs, USA, Director, Objective Force Task Force

Homeland Security Command and Control ACTD
Glenn Cooper, Assistant Technical Manager, Defense Information Systems Agency

Role of the Army Reserve in Homeland Defense
Brigadier General Gary Profit, Deputy Chief Army Reserve, Office of the Chief Army Reserve

Interagency Board for Equipment Standardization and Interoperability Working Group
Trey Gannon, Senior Research Scientist, Dartmouth College

Army Homeland Security Doctrine
Larry Heystek, Homeland Security Directorate, U.S. Army Training and Doctrine Command

Capabilities of the Department of Energy Laboratories
Frank Akers, Oak Ridge National Laboratory

Perspectives of the Department of Homeland Security
Michael Lowder, Operations Branch Chief, Response Division, Department of Homeland Security

Fusion Based Knowledge (Intelligence and Information Warfare)
Dan Kuderna, Communications and Electronics Research, Development and Engineering Center

The New Research, Development and Engineering Command
Major General John Doesburg, Commanding General, Research, Development, and Engineering Command (Provisional)

Joint Doctrine for Homeland Security
Mark L. Goracke, Headquarters, Department of the Army, G-3

Long Wave Micro-Sensor (Night Vision and Sensors)
Stuart Horn, Communications and Electronics Research, Development and Engineering Center

MOSAIC (Space and Terrestrial Communications)
Larry Muzello, Communications and Electronics Research, Development and Engineering Center

JTRS Squad Level Communications (Space and Terrestrial Communications)
Perry Hugo, Communications and Electronics Research, Development and Engineering Center

Agile Commander (Command and Control)
Charles Miller, Communications and Electronics Research, Development and Engineering Center

HLS/DaVinci (Command and Control)
Charles Miller, Communications and Electronics Research, Development and Engineering Center

Defense Collaborative Tool Suite (Command and Control)
Anthony Tom, Communications and Electronics Research, Development and Engineering Center

THIRD MEETING

September 17-19, 2003
Woods Hole, Massachusetts

Meeting objectives: Complete composition/balance/bias discussions; discuss project plan and report realization; discuss first full message draft; make additional writing assignments; confirm objectives, location, and date for the next committee meeting.

FOURTH MEETING

October 27-28, 2003
Washington, D.C.

Meeting objectives: Discuss project plan and report realization, discuss concurrence draft, and discuss review process.

Appendix C

Organizational Structure of the Army

There are several ways to describe the organizational structure of the Army. For this report, the committee considered it important to highlight the unique component structure of the Army (which includes both active and reserve soldiers) and the makeup of the operational Army and the institutional Army.

THE RESERVE COMPONENTS

The organization of the U.S. Army for the majority of the 20th century and all of the post-Cold War era has consisted of three components—the active component and the two reserve components (the Army National Guard and the United States Army Reserve). Over time the mix of this total Army construct has been adjusted to best meet U.S. needs in the global community. The downsizing and restructuring after the Cold War changed the percentage of the force mix. As of September 30, 2003, the total ready reserve (Army National Guard and Army Reserve) stood at 683,256 members (Reserve Forces Almanac, 2004). At this same point, the total active force stood at 493,536 members (Uniformed Services Almanac, 2004). The ready reserve thus constitutes 58 percent of the total Army (ready reserve plus active Army).

Few foresaw the pace and magnitude of change for the U.S. military associated with the fall of the Berlin Wall in December 1989. The size of the active Army has been reduced by over one-third since that time, yet the pace of operations throughout the world has increased significantly, and the resulting deployments have placed significant stress on the total Army. Reserve component soldiers in particular have experienced more frequent and longer deployments than they had during the Cold War era. These tours have resulted in increased pressure on them, their families, and their civilian employers. The force structure of the Army mix is being reviewed as this study is being conducted, and, while it is not the purpose of this report to make recommendations in this area, this situation is relevant because it underscores the need to use the country's assets more efficiently and to ensure that the nation is getting the best return on its investment.

The National Guard

There are 50 states and 4 territories with National Guard units. Each state's or territory's Guard has both Army and Air Force components.[1] Each Guard has both a federal and a state mission. As part of the federal force, the Army Guard augments the active Army. In its state role, the Army Guard works for the governor of the state and is frequently called on to respond to disasters. The adjutant general in each state or territory is the commander of the state Guard and he or she is directly responsible to the governor. The National Guard, because of its state mission, is usually the first of the Army components to be involved in responding to disasters.

The committee recognizes the recent reorganization of National Guard assets at the state level and below by the chief of the National Guard Bureau and believes that this will ultimately assist the Department of Defense (DOD) in its efforts to provide support to emergency responders.

No organization is better suited than the Army National Guard to provide that rapid assistance. This is not a new mission. The formation of the Guard in the 17th century was designed to protect the settlers in the New World, and the flexibility provided by its dual federal and state status, as well as the fact that there is an armory within 50 miles of 99 percent of the U.S. population, makes this reserve Army component a natural choice for emergency response assignments.

The U.S. Army Reserve

U.S. Army Reserve (USAR) elements are located in 962 locations across the United States and in U.S. territories, as well as in selected areas overseas. The force structure for the USAR is concentrated primarily in the combat service support area. USAR units provide specialized capabilities, particularly as military police and in civil matters involving civil affairs, signal communications, engineering, chemical operations, water purification, and so on. That such capabilities are very much in demand is evidenced by the fact that of the total number of soldiers who were forward deployed at the end of August 2003, more than one-

[1]This report is concerned only with the Army Guard.

third were USAR soldiers. This not only illustrates the critical role of the Army Reserve in enabling the Army to provide a full range of capabilities, but also underscores the increased reliance on reserve components as a full partner in performing many of the day-to-day missions for the U.S. Army.

Military support to civilian authorities has always been a core competency of the USAR, which has a history of performing these functions. As a strictly federal force, the Army Reserves generally require a presidential declaration of emergency in order to be used to support civilian authorities. However, once made available, the critical USAR infrastructure has proved invaluable in assisting civilian emergency responders in times of crisis. Emergency planning liaison officers have interacted with Federal Emergency Management Agency (FEMA) personnel on many disaster responses and provide an existing coordination mechanism.

In addition, the Army Reserve Network (ARNET), a network of communications systems throughout the United States, ties together the Army Reserve locations. It provides a means of transmitting voice, data, and video, and if combined with GuardNet (the parallel network of the National Guard) would most likely provide a secure backbone system of communications across the United States. A backbone system such as this, whether it uses GuardNet and ARNET or not, is absolutely critical to coordinating a national operational concept for emergency response.

OPERATIONAL ARMY

The operational Army consists primarily of tactical units organized around a divisional construct. The functional grouping and size of divisions have varied over time, with today's division consisting of a mix of combat and support units totaling between 15,000 and 20,000 soldiers. Currently the operational Army has 10 divisions in the active component, 6 divisions in the National Guard, and a variety of other nondivisional units in all three components (i.e., the active Army, the National Guard, and the U.S. Army Reserve) that provide a full range of capabilities to the nation. These capabilities in turn provide a full range of options to the National Command Authority for combating the broad spectrum of threats present in a dangerous and unpredictable world. The operational Army is made up of all three components and conducts operations as required.

INSTITUTIONAL ARMY

Less well known than the operational Army, but equally important, is the institutional Army. It consists primarily of a recruiting command to supply personnel for the force, a training base for individuals, and a wholesale logistical system that is tailored to properly equip and sustain the Army. Linking the operational Army and the institutional Army are a common doctrine[2] and a concepts-based requirement system.

CONCEPTS-BASED REQUIREMENT SYSTEM

The Army determines the capabilities necessary to accomplish its missions by means of a concepts-based requirement system. This system looks at all missions, stated or implied, and determines the concept of operations necessary for Army units to accomplish the missions. This analysis, conducted by the United States Army Training and Doctrine Command, analyzes the Family of War Plans from the various combatant commanders, various forms of guidance received from the Joint Staff and the National Command Authority, and specific guidance provided by the Army leadership. The analysis defines a specific set of broad capabilities, such as the need to deploy the force, the need to fight and win the nation's wars, the need to sustain the Army for periods of prolonged operation, and the need to reconstitute or change the Army. These broad mission areas can be broken down into units and individual requirements—which then determine the equipment, training, and sustainment packages necessary to provide ready capabilities.

Assessing Readiness

Readiness is then assessed primarily on the percentage of equipment fielded, the percentage of people available to a given unit, the status of training, and the quality of the sustainment package. A number of intangible factors are also involved in assessing readiness, and this aspect of the assessment is much more an art than a science. However, considering its recent performance in Afghanistan and Iraq, the U.S. Army has done a good job not only of measuring readiness but also of preparing its units for a broad range of capabilities.

Response to Terrorism

A modified concept-based requirement system has direct applicability to the emergency responder community. Terrorist attacks on U.S. soil have ratcheted up the level and types of capabilities required for emergency responders. No longer can it be said that being able to deal with a natural disaster such as a hurricane or tornado is sufficient; such capabilities help, but they are not enough. The com-

---

[2]Doctrine is, in football terms, the "playbook." The common language associated with this doctrine allows for the execution of planned options as well as the ability to take advantage of opportunities based on "audibles at the line of scrimmage." It has been perfected over decades and continues to serve the Army well.

mittee believes that even an all-hazards, multidisciplinary approach to training, while a gigantic step in the right direction, is not going to be enough. It will serve as a good start, but the committee suggests that scenario-driven exercises involving chemical, biological, nuclear, and high-explosive attacks be used as a means of defining initial capabilities—with particular emphasis on identifying command-and-control capabilities gaps—for the emergency responder community. Future programs of exercises could be used to refine these capabilities and to assess levels of training. Once the needed capabilities are defined, the focus should be on providing these capabilities to the desired level and then improving on them through technology as required. The committee believes that through a system similar to the Army's concept-based requirement system, the emergency responder community could greatly benefit from the work that the Army has done in harnessing the power of situation awareness in the area of command, control, communication, computers, intelligence, surveillance, and reconnaissance (C4ISR) in filling many gaps and refining current capabilities.

For example, information technology has enabled the Army to focus on answering three questions: (1) Where am I? (2) Where are my buddies? and (3) Where is the enemy? Accurately answering these questions is the primary task of the Army's C4ISR system. Being able to answer them successfully in real time also promises a true revolution in the way operations are conducted. Although affordability issues will be involved and prioritization will be required, a system similar to the concept-based requirement system of the Army would greatly improve national preparedness.

REFERENCES

Reserve Forces Almanac. 2004. 2004 Reserve Forces Almanac. Falls Church, Va.: Uniformed Services Almanac, Inc.

Uniformed Services Almanac. 2004. 2004 Uniformed Services Almanac. Falls Church, Va.: Uniformed Services Almanac, Inc.

Appendix D

Army Acquisition System

ARMY TECHNOLOGY AND SYSTEMS DEVELOPMENT

There are several phases to Army technology development, starting with a requirements analysis. In many cases, existing commercial technology can be adapted to military use—for example, computer technology. In other cases, no technology exists to meet the need, and it must be developed.

A technology development program normally flows through several phases, although a specific program can start at any phase, depending on the requirement and the maturity of any existing technology. The technology development phases are as follows: basic technology development, concept demonstration, technology demonstration, and system development.

The basic technology development, concept demonstration, and technology demonstration phases are normally called a science and technology (S&T) program. As the name implies, the focus of these programs is the science and the technology.

The systems development phases are usually called the acquisition phases, which normally encompass systems development, production, and life-cycle support of the system. The focus in these programs is primarily on engineering and support issues. Cost issues are also extremely important in the acquisition phases, with a focus on life-cycle cost (i.e., including the cost of systems engineering development, production, and maintenance and then final disposition of the system). Each phase has its unique characteristics and focus.

Requirements Analysis

The requirements analysis, the start of any program, looks at the capability needed by the user and generates a user's need statement. For example, the capability may be that a squad leader needs to know where his or her squad member is in a building in an urban environment. The material developer, working closely with the user, uses a prioritized scheme to meet that need. The first approach is to use existing equipment and a change in technique, tactics, or procedures to meet the need. If that is not possible, the next priority is to modify existing equipment to meet the need. If this is not possible, the material developer initiates a technology development program.

For a technology development program there is again a priority scheme. The first priority is to use commercial off-the-shelf technology and to modify it for military use. Again, for this example, the material developer may look at using the commercial Global Positioning System (GPS) for the individual soldier. (Unfortunately, GPS doesn't work inside buildings.) If there is no technology available in the commercial sector, the developer initiates a development program, described below.

Basic Technology Development

Basic technology development is focused on basic research toward a technology with potential application to the military user. The Army usually puts its research dollars into technologies that would not normally be developed in the private sector. For example, improved infrared semiconductors, which would eventually improve the Army's night vision capability, are good candidates for development. Likewise, stealth technology would be a candidate, as it has great military utility and limited civilian application. Contrast such technologies with computer chips or basic communications technology, for which the commercial sector drives investment that the Army should try to leverage.

Concept Demonstration

Once a technology has been developed, it can move to concept demonstration, a proof-of-principle experiment showing the potential application of the technology. Usually the technology is not completely mature at this stage and will need further development. For example, using a through-the-wall radar to show that individuals can be "seen" is a proof of principle—even though all the components are not yet mature.

Technology Demonstration

A technology demonstration, as well as an advanced technology demonstration, represents a more mature (although not completely mature) technical approach and sophisticated demonstration of military utility. The technology may be developed by the military, or it may be a commercial off-the-shelf product adapted for military use, or a mix. An even more mature advanced concept technology demonstration uses mature technology and normally includes a leave-behind system for limited military use, as well as a logistics support package for a couple of years.

Systems Development or Acquisition Programs

Once a technology demonstration has been successfully conducted, the program may move into an acquisition phase. This is the most costly part of the development cycle, and the focus shifts from technology to engineering the system for military use. The focus here is normally life-cycle management, which includes the costs of systems engineering development, production, testing, maintenance and logistics support, upgrades, and disposal. Under usual circumstances, and depending on the complexity of the system, the systems engineering phases can take from 2 to 5 years. The testing during this phase is usually very comprehensive, involving the user community to ensure that the system meets the user's needs.

OTHER DEVELOPMENT ISSUES FOR HOMELAND SECURITY CONSIDERATION

The focus of this report is technology that could be useful for emergency responders. In addition, it may be worthwhile for the homeland security user community to consider other resources that the Army has spent many years and dollars developing that support technology development. Clear examples include, in particular, the engineering support from the Research, Development, and Engineering Centers (RDECs), the testing capability resident in the Army, and the logistics and maintenance concepts.

Army Materiel Command Research and Development Command

The Army recently created the Army Materiel Command Research and Development Command to better coordinate its research and technology efforts. The command consists primarily of a headquarters element, which provides supervision and coordination; the RDECs, which provide the demonstration and acquisition support to the Army; and the Army Research Laboratory, which is focused on basic technology development and early concept demonstration.

- *Army Research Laboratory.* The Army Research Laboratory, headquartered in Adelphi, Maryland, is the responsible activity within the Army for early technology development, including basic research and early concept development. It has a talented technical pool, supporting primarily internal research of technologies of interest to the Army. For example, the laboratory's Materials Division works on new, sophisticated materials for future infrared systems and has extensive capability in automatic target detection, acoustics, and so on. The parallel organization in the Army that supports external funding is the Army Research Office, with an emphasis on basic research.

- *RDECs.* The Army has several RDECs with a wealth of scientific and engineering talent that can be called on by the homeland security community. These centers can support buying decisions with technical evaluations, can do component and system evaluation to support Qualified Product Lists, and can help develop technology demonstrations. Currently, the Communication and Electronic Command RDEC is supporting the New York City Transit Authority in a demonstration project.

Testing

The Army, as well as other components of the DOD, has significant testing ranges as well as a system for testing equipment to ensure that it meets users' needs. The testing includes operation or performance testing as well as maintenance and logistics support testing.

Logistics and Maintenance Concepts

Critical to any successful system for use by either the military or emergency responders is a logistics and maintenance concept. Usually developed during the systems development step/phase, the maintenance and logistics concept supports the sustainability of an item of equipment for many years, ensuring that the equipment can be serviced, repaired as necessary, replaced, and disposed of at the end of its life cycle. The Army has many years of experience in this area, usually resident in its logistical and commodities centers.

Appendix E

C4ISR Capabilities for the Future Force

Some of the anticipated operational capabilities to be afforded to the Future Force by C4ISR are listed below. Consistent with the rest of the report, this appendix is organized in the following groupings: command, control, and computers (C3); communications (C); and intelligence, surveillance, and reconnaissance (ISR). The sections for C3 and communications are further divided under the subheadings *"See First,"* *"Understand First,"* and *"Act First"* and augmented by another important parameter, *"Ensure Reliability."* Each capability is listed under the parameter that fits it best; however, it is noted that these capabilities usually support other parameters and components to some extent as well.

ANTICIPATED OPERATIONAL CAPABILITIES DERIVED FROM COMMAND, CONTROL, AND COMPUTERS

See First

- Commanders will have access to a common operational picture (COP) with timely updates to ensure near-perfect situational awareness and to overcome the fog of war.
- Information will be automatically "pushed" to the commander as well as being "pulled" from the network in accord with immediate needs.
- The Future Force Warrior (FFW) will use forward sensor fusion and ubiquitous, assured network communications to enable improved battle management, command and control, and situational awareness.

Understand First

- Information will be fused at the commander's level to avoid information overload and enable complete situational understanding.
- Automated event-tracking capability will alert commanders to deviations from the operations plan or to unanticipated exigencies.
- Commanders will be provided state-of-the-art collaborative, distributed, real-time decision aids to facilitate informed decisions.
- Computer systems will be designed to enable individual soldiers to be recognized by any system and to be uniquely identified with appropriate rank, priority, and information needs.
- The FFW will be integrated, interoperable, and interfaced with the unit of action systems and is capable of independent operations with joint assets and firepower.
- The FFW will have command and control of organic tactical mobile robots and can interface with Future Combat Systems (FCS) robotic platforms.

Act First

- The operations plan will be continuously updated with inputs from higher and lower echelons as events unfold.
- The command post will be wherever the commander is located, be that location mounted, dismounted, or airborne.
- Changes in leadership on the battlefield will be automatically accommodated on the fly to ensure continuity of command.
- Automated synchronization of maneuver, firepower, and reconnaissance, surveillance, and target acquisition (RSTA) will be provided.
- Ad hoc sensor-command-shooter links will be automatically or semi-automatically established to maximize effective firepower.

Ensure Reliability

- Computer hardware will be robust and rugged for operations in the field.
- Computer system components will be modular for easy and rapid repairs in the field or for upgrades.
- Computers will be protected against hostile penetration with advanced firewalls and other security systems.
- Software applications will be robust and flexible to accommodate interruptions in network services and changes in data rates without crashing.
- Software will be readily serviced or upgraded in the field via the information network, without on-site technicians.

ANTICIPATED OPERATIONAL CAPABILITIES DERIVED FROM COMMUNICATIONS

See First

- The network will utilize ground, airborne, and space communication line-of-sight and non-line-of-sight links to achieve continuous, uninterrupted connectivity on the move.
- Networks will provide continuous position and navigation functions of all blue (friendly) entities with minimum self-disclosure to enemy.
- Entities (including individual soldiers) will be connected to the network to enable each warfighter to have access to needed information and to enable each system/sensor to contribute relevant data that it may collect to the knowledge pool.

Understand First

- The network will facilitate battlefield Identification: Friend or Foe by providing local situational awareness to engaged tactical units.
- The network will provide reach-back through the global information grid (GIG) for national source information, as well as providing administrative and logistic support for all combat support and combat service functions.
- Connectivity will be provided to local military or civilian networks as required.
- The network will be backward-compatible to extend to legacy systems.
- Above all, the communications system must be robust to ensure that critical information is provided to the warfighters when needed.
- Network management will be essentially built in, requiring little to no on-site support. Upgrades or servicing can be done remotely via the network itself.

Act First

- The network will be ad hoc, that is, self-configuring, allowing entities and nodes to enter and leave automatically without operator involvement.
- The network will make maximum use of all available spectral bandwidth by dynamically adapting to battlefield exigencies and the commander's priorities.
- Network capabilities will allow voice, data, and video in accord with the commander's intent and priorities and the battlefield situation.
- The network is tied into the GIG to enable variable joint and coalition connectivity and operations.

- Network protocols will facilitate multiple levels of security.
- Network protocols will accommodate temporary interruptions in connectivity (for example, if a vehicle passes under a bridge) without requiring resetting.

Ensure Reliability

- The network will be robust against environmental effects such as rain, fog, foliage, buildings, and structures.
- The network will be robust against jamming countermeasures, taking automatic counter-countermeasures to minimize adverse effects.
- Networks will be assured as well as secure, providing protection against insertion of specious data and denying access to hostile or unauthorized personnel or forces.
- To the extent possible, the network will preclude interception and network analysis by hostile forces.
- The network hardware and protocols will be commercially based to the maximum extent possible in order to facilitate technology insertions as they evolve.

ANTICIPATED OPERATIONAL CAPABILITIES DERIVED FROM INTELLIGENCE, SURVEILLANCE, AND RECONNAISSANCE

See First

- Manned and unmanned ground, air, and space systems will extend vision beyond the line of sight to provide continuous, ubiquitous battlefield monitoring through both passive and aggressive RSTA.
- Sensors will be available to see through walls in urban operations.
- The ISR system will employ the full range of operational variables—terrain; weather; friendly and enemy forces; and noncombatants—and detect threat actions in all environments.
- The ISR system will manage the overall application of organic sensor assets in accord with the commander's intent and needs.
- The ISR system will provide standoff means to detect mines, booby traps, and command-detonated munitions "in stride" so as to maintain operational tempo.
- Semiautomated pattern analysis will be performed to detect, locate, and identify enemy combatants and systems.
- Control of sensors and information collection, as well as analysis, will be distributed via the network to eliminate single-point vulnerability.

- The sensor system will be designed to operate in all weather and all terrain, against enemy entities that are dispersed, covered and concealed, masked, and fleeting.
- Joint combat identification measures will be integrated.

Understand First

- Sensor data collected from both manned and unattended sensor networks will be processed, networked, and fused into an integrated COP for unprecedented situational awareness and understanding.
- Commanders will be able automatically or with software decision aids to sort out from a variety of enemy data entries which are most dangerous and which have higher payoffs for engagement at tactical standoff.
- Highly precise data on targets will flow from sensor to shooter and enable reliable and timely battlefield damage assessment.
- Joint, Army, and coalition manned and unmanned air, ground, and space RSTA assets will be used synergistically to gain and maintain contact with enemy elements and to provide high-resolution combat information on terrain and weather.
- Near-real-time friend, foe, or noncombatant identification across the spectrum of operations will be achieved through platform-to-platform, platform-to-soldier, soldier-to-platform, and soldier-to-soldier interrogation.
- Means will be provided to sort through decoys, deception, and dis-information.
- Robotic systems will be employed for certain high-risk situations.
- Means will be provided to defeat the enemy's ISR systems through the use of obscurants, jamming, signature reduction, deception, and pattern avoidance techniques in order to see, understand, and act first.

Appendix F

C4ISR Capabilities for Civilian Emergency Responders

Some of the operational capabilities that could be afforded civilian emergency responders by command, control, communications, computers, intelligence, surveillance, and reconnaissance (C4ISR) are listed below. Consistent with the rest of the report, this appendix is organized in the following groupings: command, control, and computers (C3); communications (C), and intelligence, surveillance, and reconnaissance (ISR).

COMMAND, CONTROL, AND COMPUTERS

Following are C3 operational capabilities that could be afforded to emergency responders:

- A means to conduct realistic, high-quality training programs and exercises that employ operational systems, with embedded mission rehearsal, simulations, and distance education. In particular, modeling and simulations must be capable of modeling the dispersal and effects of chemical, biological, and radiological agents, as well as blast effects in complex urban terrain and the interior of buildings. Training and exercises must be scalable, to include different types of emergency responders, jurisdictions, and levels of government and, where possible, "turn-key" operations.
- An ability to continuously monitor high-value targets and critical emergency responder infrastructure and to communicate status, whenever needed, particularly in urban centers, of the interior of large buildings and underground facilities. Monitoring systems would employ automatic alarms.

- Means to identify and recognize threat-relevant information, analyze data, and present the information so that it can be assessed and understood.
- Enhanced classification and mitigation capacity for command-and-control centers, including the ability to integrate sensor data with symptoms and pathology in order to classify medical threats and to provide mitigation guidelines for both protecting emergency responders and providing immediate treatment for victims.
- The ability to know and visualize the location of an attack in three dimensions and to track in real time the location and status of all emergency responders.
- Means to identify, establish, manage, and control security perimeters and to manage the flow of traffic in and out of a disaster area, in particular for responses to chemical and biological attacks. Perimeters should be capable of being established within minutes by the first on-scene emergency responders and capable of being rapidly modified as required.
- Automated support for handling large numbers of casualties and the capacity to share information with first responders, medical personnel, and public health officials in a manner that both facilitates care and respects individual patient rights.
- Means to rapidly collect and disseminate information on as many as thousands of missing persons to emergency responders and law enforcement personnel, providing the means to respect patient rights and reunite missing persons with their families.
- The ability to determine lists of supplies required for responses to any kind of large-scale disaster or terrorist attack.
- Means to manage logistical inventories, including delivery of supplies and support equipment on demand, provision for the rapid use of available supplies based on current needs, development of usage trends, and projections of future demands for responding to a large-scale terrorist attack.
- The ability to coordinate among law enforcement, medical personnel, medical examiners/coroners, and veterinary and public health officials for epidemiological surveillance information for attributing the source of a biological attack.
- The ability to manage relocation destinations and shelters for evacuees following a large-scale terrorist attack.
- Means to manage volunteer personnel, equipment, and supplies for dealing with the consequences of a large-scale terrorist attack.
- The ability to manage traffic in an evacuation or in and around a large chemical, biological, or radiological disaster site, to include monitoring traffic flows, routes, and destinations, as well as traffic accidents and other traffic blockages.

COMMUNICATIONS

Following are operational communications capabilities that could be afforded to civilian emergency responders:

- Means to seamlessly connect and integrate multiple interagency users and information and communications systems. The system must be scalable to include different types of emergency responders, jurisdictions, and levels of government and to accommodate the different types of information that might be required (e.g., audio, video, or data).
- Means to provide information assurance that is scalable to guarantee the availability, security, and integrity of information required by emergency responders for different types of operations. The system must be capable of operating in complex urban terrain and provide redundancy in the case of loss of critical infrastructure.
- Scalable, interoperable, on-demand communications between on-scene emergency responders.
- Reliable communications link between on-scene detectors and command centers, allowing alerting and subclinical information to be integrated without operator intervention.
- Means to transfer information on emergency responder status to an off-site command post or monitoring station. Preferably this system would be integrated into emergency responder gear, add little additional weight, have an independent power source, and require no support or attention from individual emergency responders.
- Communications equipment that is integrated into personal protective equipment, allows hands-free operations and use with gloves, is easy to train on and use, and requires little logistical support (e.g., batteries).

INTELLIGENCE, SURVEILLANCE, AND RECONNAISSANCE

Following are operational ISR capabilities that could be afforded to civilian emergency responders:

- Means to conduct intelligence preparation for operations by identifying what threat and critical infrastructure data need to be disseminated and who needs to receive the information, and to deliver the information to the appropriate user at the required level of security classification.
- On-scene detection capabilities, including the capacity to detect suspicious objects, secondary devices, and the post-attack location of agents and down-wind hazards. These capabilities would need to be accurate, reliable, and rapid; require minimal logistical support (e.g., power requirements); and be both human-portable and vehicle-mounted. Detection

systems should be capable of determining biological, chemical, radiological, and explosive hazards.

- Means to provide the three-dimensional location of emergency responders and to monitor their physical and physiological status.
- The capacity to assess radiological, chemical, and biological threats from outside the danger area at a safe distance and to rapidly analyze and disseminate the information.
- The ability to rapidly interrogate sensors monitoring critical infrastructure and target areas, integrate data, and provide mitigation guidelines.
- The ability to collect and rapidly disseminate data on weather and environmental conditions (e.g., winds, temperature, humidity, air quality) in order to support modeling of weapons effects and provide information to emergency responders so that they can avoid threats, mitigate risks, and adjust containment areas. Weather support capabilities would include the ability to assess the environmental impact on interior and exterior microclimates (e.g., inside buildings) and account for complex urban terrain and building effects.
- The ability to assess threats inside buildings and in underground infrastructure; to identify and distinguish emergency responders, victims, and perpetrators; to evaluate risks; and to determine location and status of perpetrators, hazardous devices, and weapons.
- The ability to provide early detection, identification, assessment, and tracking of exposure to biological agents through epidemiological and veterinary surveillance.
- A capacity to run rapid field tests of agriculture, livestock, and pets to identify and assess biological, chemical, and radiological threats.
- The ability to provide rapid assessments of the integrity of structures in the wake of explosions and fires to on-site emergency responders.
- The ability to rapidly locate and assess injured/contaminated victims in a chemical, biological, or radiological environment in areas with and without structural collapse.
- The means to detect physical threats against emergency responders (such as from snipers, mines, mortars, and shoulder-fired weapons).
- The capacity to detect nonlethal attacks such as electronic jamming on emergency responders and responder assets.

Appendix G

Criteria for Technology Readiness Levels

TABLE G-1 Criteria for Technology Readiness Levels (TRLs)

TRL	Task Accomplished	Description
1	Basic principles observed and reported	Lowest level of technology readiness. Scientific research begins to be translated into applied research and development. Examples might include paper studies of a technology's basic properties.
2	Technology concept or application formulated	Invention begins. Once basic principles are observed, practical applications can be invented. The application is speculative, and there is no proof or detailed analysis to support the assumption. Examples are still limited to paper studies.
3	Analytical and experimental critical function or characteristics proof of concept	Active research and development are initiated. These include analytical studies and laboratory studies to physically validate analytical predictions of separate elements of the technology. Examples include components that are not yet integrated or representative.
4	Component or breadboard validation in laboratory environment	Basic technology components are integrated to establish that the pieces will work together. This is relatively "low fidelity" compared with the eventual system. Examples include integration of ad hoc hardware in a laboratory.

TABLE G-1 Continued

TRL	Task Accomplished	Description
5	Component or breadboard validation in a relevant environment	Fidelity of breadboard technology increases significantly. The basic technological components are integrated with reasonably realistic supporting elements so that the technology can be tested in a simulated environment. Examples include high-fidelity laboratory integration of components.
6	System/subsystem model or prototype demonstration in a relevant environment	Representative model or prototype system, which is well beyond the breadboard tested for TRL 5, is tested in a relevant environment. This level represents a major step up in a technology's demonstrated readiness. Examples include testing a prototype in a high-fidelity laboratory environment or in a simulated operational environment.
7	System prototype demonstration in an operational environment	Prototype near or at planned operational system. This level represents a major step up from TRL 6, requiring the demonstration of an actual system prototype in an operational environment, such as in an aircraft, vehicle, or space. Examples include testing the prototype in a testbed aircraft.
8	Actual system completed and flight-qualified through test and demonstration	Technology has been proven to work in its final form and under expected conditions. In almost all cases, this TRL represents the end of true system development. Examples include developmental test and evaluation of the system in its intended weapon system to determine if it meets design specifications.
9	Actual system flight proven through successful mission operations	Actual application of the technology in its final form and under mission conditions, such as those encountered in operational test and evaluation. In almost all cases, this is the end of the last bug-fixing aspects of true system development. Examples include using the system under operational mission conditions.

SOURCE: Adapted from: U.S. Army. 2001. Army Science and Technology Master Plan. Washington, D.C.: U.S. Army Office of the Deputy Assistant Secretary of Defense for Research and Technology (Chief Scientist).